IQ CALIBRATION TECHNIQUES FOR CMOS RADIO TRANSCEIVERS

ANALOG CIRCUITS AND SIGNAL PROCESSING SERIES

Consulting Editor*: Mohammed Ismail. *Ohio State University

Related Titles:

CMOS CURRENT-MODE CIRCUITS FOR DATA COMMUNICATIONS
Yuan, Fei
ISBN: 0-387-29758-8

ADVANCED DESIGN TECHNIQUES FOR RF POWER AMPLIFIERS
Rudiakova, A.N., Krizhanovski, V.
ISBN 1-4020-4638-3

CMOS CASCADE SIGMA-DELTA MODULATORS FOR SENSORS AND TELECOM
del Río, R., Medeiro, F., Pérez-Verdú, B., de la Rosa, J.M., Rodríguez-Vázquez, A.
ISBN 1-4020-4775-4

Titles in former series International Series in Engineering and Computer Science:

SIGMA DELTA A/D CONVERSION FOR SIGNAL CONDITIONING
Philips, K., van Roermund, A.H.M.
Vol. 874, ISBN 1-4020-4679-0

CALIBRATION TECHNIQUES IN NYQUIST A/D CONVERTERS
van der Ploeg, H., Nauta, B.
Vol. 873, ISBN 1-4020-4634-0

ADAPTIVE TECHNIQUES FOR MIXED SIGNAL SYSTEM ON CHIP
Fayed, A., Ismail, M.
Vol. 872, ISBN 0-387-32154-3

WIDE-BANDWIDTH HIGH-DYNAMIC RANGE D/A CONVERTERS
Doris, Konstantinos, van Roermund, Arthur, Leenaerts, Domine
Vol. 871 ISBN: 0-387-30415-0

METHODOLOGY FOR THE DIGITAL CALIBRATION OF ANALOG CIRCUITS AND SYSTEMS: WITH CASE STUDIES
Pastre, Marc, Kayal, Maher
Vol. 870, ISBN: 1-4020-4252-3

HIGH-SPEED PHOTODIODES IN STANDARD CMOS TECHNOLOGY
Radovanovic, Sasa, Annema, Anne-Johan, Nauta, Bram
Vol. 869, ISBN: 0-387-28591-1

LOW-POWER LOW-VOLTAGE SIGMA-DELTA MODULATORS IN NANOMETER CMOS
Yao, Libin, Steyaert, Michiel, Sansen, Willy
Vol. 868, ISBN: 1-4020-4139-X

DESIGN OF VERY HIGH-FREQUENCY MULTIRATE SWITCHED-CAPACITOR CIRCUITS
U, Seng Pan, Martins, Rui Paulo, Epifânio da Franca, José
Vol. 867, ISBN: 0-387-26121-4

DYNAMIC CHARACTERISATION OF ANALOGUE-TO-DIGITAL CONVERTERS
Dallet, Dominique; Machado da Silva, José (Eds.)
Vol. 860, ISBN: 0-387-25902-3

ANALOG DESIGN ESSENTIALS
Sansen, Willy
Vol. 859, ISBN: 0-387-25746-2

DESIGN OF WIRELESS AUTONOMOUS DATALOGGER IC'S
Claes and Sansen
Vol. 854, ISBN: 1-4020-3208-0

MATCHING PROPERTIES OF DEEP SUB-MICRON MOS TRANSISTORS
Croon, Sansen, Maes
Vol. 851, ISBN: 0-387-24314-3

LNA-ESD CO-DESIGN FOR FULLY INTEGRATED CMOS WIRELESS RECEIVERS
Leroux and Steyaert
Vol. 843, ISBN: 1-4020-3190-4

SYSTEMATIC MODELING AND ANALYSIS OF TELECOM FRONTENDS AND THEIR BUILDING BLOCKS
Vanassche, Gielen, Sansen
Vol. 842, ISBN: 1-4020-3173-4

LOW-POWER DEEP SUB-MICRON CMOS LOGIC SUB-THRESHOLD CURRENT REDUCTION
van der Meer, van Staveren, van Roermund
Vol. 841, ISBN: 1-4020-2848-2

WIDEBAND LOW NOISE AMPLIFIERS EXPLOITING THERMAL NOISE CANCELLATION
Bruccoleri, Klumperink, Nauta
Vol. 840, ISBN: 1-4020-3187-4

IQ CALIBRATION TECHNIQUES FOR CMOS RADIO TRANSCEIVERS

by

Sao-Jie Chen

National Taiwan University, Taipei, Taiwan

and

Yong-Hsiang Hsieh

Muchip, Hsin-Chu, Taiwan

Springer

A C.I.P. Catalogue record for this book is available from the Library of Congress.

ISBN-13 978-90-481-7279-5 e-ISBN-13 978-1-4020-5083-1

Published by Springer,
P.O. Box 17, 3300 AA Dordrecht, The Netherlands.

www.springer.com

Printed on acid-free paper

Contents

List of Figures

List of Tables

List of Abbreviations

A/D	Analog-to-Digital Converter
AFC	Automatic Frequency Control
AGC	Auto Gain Control
AP	Access Point
BB	Baseband
BW	Bandwidth
CCK	Complementary Code Keying
DAC	Digital-to-Analog Converter
DSSS	Direct Sequence Spread Spectrum
ETSI	European Telecommunications Standards Institute
EVM	Error Vector Amplitude
FE	Front-End
FSK	Frequency-Shift Keying
HPF	High Pass Filter
I	In-phase
IF	Intermediate Frequency
IL	Insertion Loss
LAN	Local Area Network
LNA	Low Noise Amplifier
LO	Local Oscillator
LPF	Low Pass Filter
OFDM	Orthogonal Frequency Division Multiplexing
PA	Power Amplifier
PAPR	Peak to Average Power Ratio
PCB	Print Circuit Board
PER	Package Error Ratio

PLL	Phase Lock Loop
Q	Quadrature
QAM	Quadrature Amplitude Modulation
QPSK	Quadrature Phase Shift Keying
RF	Radio Frequency
RX	Receiver
S/C	Sample and Compare
SAW	Surface Acoustic Wave
SNR	Signal-to-Noise Ratio
SoC	System-on-Chip
SSB	Single Side Band
SWAP	Semantic Web Application Platform
TDD	Time-Division-Duplex
TX	Transmitter
VCO	Voltage Control Oscillator
VGA	Variable Gain control Amplifier
WiFi	Wireless Fidelity
WLIF	Wireless LAN Interoperability Forum

Preface

In the market of wireless communication, high data-rate transmission and high spectral efficiency have been the trend. The IEEE 802.11 a/g standards working at 5GHz/2.4GHz ISM bands can support data rates up to 54Mbits/s using OFDM modulation. The newly proposed 802.11n technology now uses 64-QAM to achieve higher spectral efficiency. The DVB and many other systems will also use QAM for its data transmission.

The cost of achieving this higher spectral efficiency using higher order QAM is that the transmitter and receiver requires a higher signal to noise ratio (SNR) in order to modulate and demodulate the signal with the same level of error rate performance (relative to a baseline BPSK or QPSK system). The dominant vectors on SNR degradation are noise floor, signal distortion, down conversion (up conversion for transmitter) local phase noise and the modulator/demodulator I/Q gains and phases imbalance.

There are a lot of vectors that degrade the matching of gains and phases between I/Q signals: the instinct layout mismatch, the random mismatch of the devices, the different temperature over the I/Q signal paths, etc. Solutions on I/Q gains and phases mismatch compensation can be classified into three schools. The first school uses a fully-digital compensation technique. For example, Eberle *et al.* [5] presented a digital compensation architecture for the I/Q mismatch problem that occurred in a digital receiver. The second one uses baseband plus RF front-end to calibrate the gain and phase mismatch. For example, baseband signal processing techniques have been used to generate a test signal to the RF front-end [6, 7]. Since the output of the RF front-end running this test signal will give information on the I/Q mismatch, the baseband can sense this output to determine a control code for the front-end circuits to compensate the I/Q mismatch. But most of the academic institutes and even business have limited man-power to develop an RF front-end plus a baseband IC for compensation purposes. The third one uses a fully-analog compensation technique without baseband circuitry to control the calibration process. For example, Hsieh *et al.* [9] first presented an auto-I/Q calibrated transceiver with no more than 5% hardware overhead for the calibration circuitry. This book will use an 802.11g transceiver design as an example to give a detail description on the I/Q gains and phases imbalance auto-calibration mechanism.

The first part of this work discusses the reasons on why we would like to design an 802.11g transceiver and the architecture of the transceiver. We decide to use a superheterodyne architecture for the transceiver with the baseband filters and to re-use the IF Variable Gain Control (VGA) in both modulator and demodulator to reduce the chip size. System link budget calculation has also been showed on this part.

The second part include Chapters 4 and 5. The DC offset cancellation in modulator, and the I/Q gains and phases imbalance auto-calibration in both modulator and demodulator are discussed on this part. The final chapters illustrate the chip measured result.

Sao-Jie Chen
Graduate Institute of Electronics Engineering and
Department of Electrical Engineering
National Taiwan University
Taipei, Taiwan, 2006

Yong-Hsiang Hsieh 謝永強
RF IC Product Design Department
Muchip
Hsin-Chu, Taiwan, 2006

Acknowledgments

Many persons contributed to development of this book. First, Dr Syed K. Enam, and Dr David-J Chen are acknowledged for their valuable inputs to the circuits and architecture. The circuits described in the text were realized in close cooperation with the graduate students from Prof. Sao-Jie Chen's research group and with colleagues from Muchip. In particular, we want to express our appreciation to the following persons: Dennis Cheng, Wei-Yi Hu, Shin-Ming Lin, Chao-Liang Chen, Wen-Kai Li for their contributions to the work described in the text. Finally, we would like to thank our parents and our wives.

Sao-Jie Chen
Graduate Institute of Electronics Engineering and
Department of Electrical Engineering
National Taiwan University
Taipei, Taiwan, 2006

Yong-Hsiang Hsieh 謝永強
RF IC Product Design Department
Muchip
Hsin-Chu, Taiwan, 2006

Chapter 1

INTRODUCTION

This chapter will present the standards of different wireless applications and describe the development of wireless products in the 21st century. Then more detailed and comparative explanation on a series of IEEE 802.11 standards will be shown. Since IEEE 802.11g is the best choice considering performance and cost now in the 802.11x series, the difficulties faced on IEEE 802.11g transceiver design will be discussed, especially the I/Q mismatch in quadrature modulator and demodulator.

1. WIRELESS LAN STANDARDS

For wireless LAN communications, many different standards were published by different organizations worldwide. This section gives an overview on the most popular standards.

1.1 IEEE 802.11

The main problem of radio network acceptance in the market place is that there is not one unique standard like Ethernet with a guaranteed compatibility between all devices, and the proprietary standards pushed by each independent vendor are incompatible between themselves. But corporate customers like to have an established unique standard, and thank to the effort of vendors who have joined the IEEE standard for radio LANs, the IEEE 802.11 has been created.

1.2 HiperLan

The development of HiperLan is totally different to 802.11. This standard has been designed by a committee of researchers within the European Telecommunications Standards Institute (ETSI) without strong vendors influence, and is quite different from existing products. The standard is quite simple, which owns some advanced features and has already been ratified a while ago (since summer 96).

1.3 HiperLan II

HiperLan II is again a standard different to HiperLan. The first HiperLan was designed to build ad-hoc networks, HiperLan II was designed for managing infrastructure and wireless distribution systems. The only similarity is that both standards are specified by the ETSI Broadband Radio Access Network group, operate at the 5GHz (5.4 to 5.7GHz) band dedicated in Europe.

1.4 OpenAir

OpenAir is a proprietary proctocal from Proxim [1]. As Proxim is one of the largest Wireless LAN manufacturers, they are trying to push OpenAir as an alternative to 802.11 through the Wireless LAN Interoperability Forum (WLIF). Proxim is the only one having all the detailed information on OpenAir.

1.5 HomeRF and SWAP

The HomeRF is a group of big companies formed from different background to push the applications of Wireless LAN in home and in small offices. This group is developing and promoting a new Radio Lan standard: Semantic Web Application Platform (SWAP).

1.6 BlueTooth

BlueTooth should not be included in the wireless LAN clan, but people keep thinking that BlueTooth is a Wireless LAN. BlueTooth is a cable replacement technology mostly developed and promoted by Ericsson with the help of Intel, offering point to point links and no native support for IP. It may be good for some applications, but not for Wireless LANs.

2. WIRELESS IN THE 21ST CENTURY

Since the advent of IEEE 802.11 wireless LAN in 1997, it has already begun to firmly establish itself as an impressive technology. Corporations, government facilities, academic institutions and traditional home users have all begun to discover ways that wireless technology can help to free them from the confines of their homes and the boundaries of their environment. Since the inception of widespread internet access information has become more available to more people than ever before. Through the use of Wireless Fidelity (WiFi) networking technology, systems will be integrated and conjoined with spectacular ease.

The future of wireless technology is limited only by the resourcefulness and creativity of those who choose to integrate it into their world. Wireless classrooms can allow students to work more freely and outside the boundaries of a traditional classroom while still able to access a world of information. Government and mobile military installations will be able to more effectively manage their personnel and security communications. Corporations are rapidly switching to WiFi networks as it not only reduces their intradepartmental paperwork and workload, but also allows them to serve their customers and employees in "real-time," a talent which is virtually beyond financial compensation. The possibilities with wireless technology are limitless and beyond the scope of our current technological conventions. With only a few years of active utilization IEEE 802.11, wireless LAN networking has already begun to spread and will most likely continue to stay spreading with incredible speed.

In all different standards, IEEE 802.11 is used most extensively. A lot of other different standards, not being used, are already being eliminated slowly. Discussion on IEEE 802.11 standard will be detailed in Section 1.3.

3. THE 802 STANDARD AND THE IEEE

If any one technical organization stands tall above the rest, it is the IEEE. It is this body of academics and technology professionals who helps to carve the path through the often precarious and cumbersome world of emerging high technology. The engineers in the IEEE have been instrumental in helping to adopt and refine protocol and operational standards for countless forms of computer and communication technology. Typically they achieve such great result by forming working groups, whose sole mission is to resolve a particular standard issue. Culminating almost ten years of discussion to deliberate the IEEE finally adopted 802 standard as their official ground level networking standard in 1990. From there on 802 continued to grow and develop into various standard specifications such as the 802.3 for Ethernet networking, and 802.11 for wireless networking in 1997.

After its initial validation by the IEEE, the 802.11 wireless networking standard quietly exploded onto the market in many different capacities, catching many IT departments by surprise. Over the last couple years, the 802.11 standard itself has begun to support various incremental enhancements and adaptations to the protocol as it grows to meet the industry's needs. Subgroups have begun to spring up, each trying to cope with a different facet or improvement to the original standard. Such revisions include 802.11a, 802.11b and 802.11g to name just a few of the major enhancements and revisions.

3.1 IEEE 802.11b

Ratified by the IEEE in July 1999, 802.11b extends the original IEEE 802.11 direct sequence spread spectrum (DSSS) standard to operate up to 11 Mbps in the 2.4-GHz unlicensed spectrum using complementary code keying (CCK) modulation. The four data rates of 1, 2, 5.5, and 11 Mbps are specified on up to three non-overlapping channels, and the lowest two rates are also allowed on up to 13 overlapping channels.

3.2 IEEE 802.11a

Ratified by the IEEE at the same time as 802.11b, the IEEE 802.11a (802.11a) standard operates in the 5-GHz spectrum. The 802.11a standard was designed for higher bandwidth applications than 802.11b, and includes data rates of 6, 9, 12, 18, 24, 36, 48, and 54 Mbps using orthogonal frequency division multiplexing (OFDM) modulation on up to 12 discrete channels.

3.3 IEEE 802.11g

In July 1999, the 802.11g [2] subcommittee was tasked to extend the 2.4-GHz unlicensed spectrum to data rates faster than 20 Mbps. The resulting 802.11g standard was ratified in June 2003. The 802.11g standard provides optional data rates of up to 54 Mbps, and requires backward compatibility with 802.11b devices to protect the substantial investments in today's

WLAN installations. The 802.11g standard includes mandatory and optional components. It specifies OFDM (the same technology used in 802.11a) and CCK as the mandatory modulation schemes with 24 Mbps as the maximum mandatory data rate, but it also provides for optional higher data rates of 36, 48, and 54 Mbps.

Table 1-1. IEEE 802.11 Specifications

	802.11b	802.11a	802.11g
Stand approved	July 1999	July 1999	June 2003
Maximum data rate	11Mbps	54Mbps	54Mbps
Modulation	CCK	OFDM	OFDM and CCK
Data Rates	1, 2, 5.5, 11Mbps	6, 9, 12, 18, 24, 36, 48, 54Mbps	CCK: 1, 2, 5.5, 11 OFDM: 6, 9, 12, 18, 24, 36, 48, 54Mbps
Frequency	2.4 – 2.497GHz	5.15 – 5.35GHz 5.425 – 5.675GHz 5.725 – 5.875GHz	2.4 – 2.497GHz

3.4 Performance and Characteristic

Data rate, range, throughput, and compatibility vary among the three WLAN standards. These variations are caused by differences in frequency, modulation schemes, and number of data rates.

As distance from the access point increases, 802.11 based products provide reduced data rates to maintain connectivity. The 802.11g standard has the same propagation characteristic as 802.11b, because it transmits in the identical 2.4-GHz frequency band. Some 802.11b and 802.11g products share the same propagation characteristics, their implementations provide roughly the same maximum range at the same data rate. Moreover, the 5-GHz radio signals do not propagate as well as the 2.4-GHz radio signals, the product range of 802.11a is limited compared to the 802.11b or 802.11g. The following figure illustrates the expected data rate of each technology at different ranges.

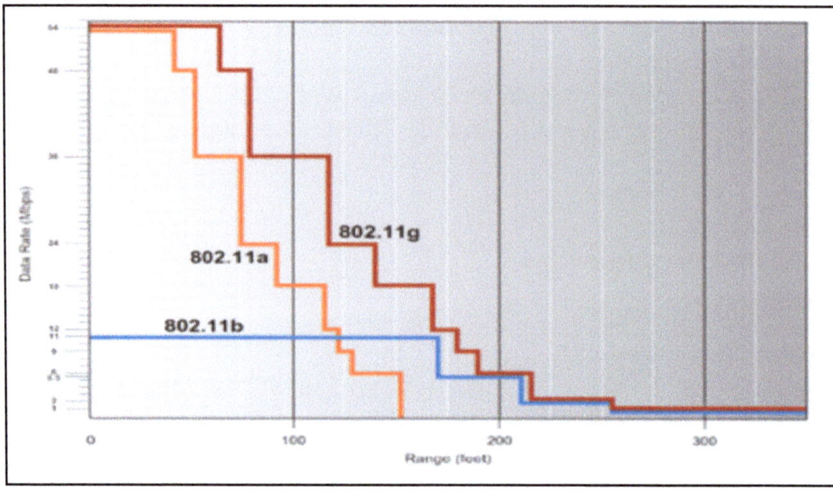

Figure 1-1. Expected 802.11a, 802.11b and 802.11g Data Rates at Varying Distance from Access Point [24].

With any network technology, there are tradeoffs between performance and cost. With wireless technologies, other factors such as range and capacity should also be considered. Table 1.2 summarizes typical user requirements and the characteristics of WLAN technology. In most cases, 802.11g provides the correct mix of characteristics for the different categories of user requirements. Upgrading to 802.11g is the easiest and least expensive choice, because it can be done gradually, without sacrificing the current WLAN infrastructure. The 802.11g Access Point (AP) automatically support existing 802.11b clients while providing increased speed to 802.11g clients.

Table 1-2. WLAN User Requirements and Technology Characteristics in the U.S. [2]

Typical WLAN User Requirements				
Type of WLAN	Peak Speed	Capacity	Range	802.11b Compatible
Enterprise	High	Variable	Variable	Yes
Public Access	Low	Medium	High	Yes
Small Business	Medium	Medium	High	Yes
Home	Medium	Low	High	Yes

WLAN Technology Characteristics				
Type of WLAN	Peak Speed	Capacity	Range	802.11b Compatible
802.11b	Medium	Low	High	Yes
802.11a	High	High	Low	No
802.11g	High	Medium	High	Yes

4. BACKGROUND AND MOTIVATION

As seen from the discussion of the foregoing paragraphs, the IEEE 802.11g provides the correct mix of characteristics for the different categories of user requirements. Because of market need, we determine to design a 802.11g transceiver

This book describes the design and implementation of an auto-I/Q calibrated CMOS transceiver for 802.11g. As shown in Table 1.1, IEEE 80211g uses an OFDM 64 QAM modulation in the maximum data transmission mode. OFDM uses a set of subcarrier frequencies, and these frequencies are orthogonal. Each subcarrier is modulated individually, the bit rate and signal strength of each subcarrier can be adapted to get maximum performance of the system (by putting more bits on the good subcarriers and less on the bad ones). Then the system splits the bits to transmit between the subcarriers, each subcarrier is modulated and then combined to produce the transmited signal (using a Fast Fourrier Transform)

The spectral efficiency of the 802.11g standard comes at the expense of a more complicated transceiver with strict requirements on the radio performance. For example, the use of 64-QAM modulation requires a signal-to-noise ratio (SNR) of 25 dB, which is substantially greater than that required by the FSK modulation in Bluetooth and the QPSK modulation in 802.11b. This high SNR translates to stringent phase noise requirements for the frequency synthesizer, tight matching constraints for both the transmitter and receiver, good circuits linearity performance and system noise performance.

5. IEEE 802.11G RF TRANSCEIVER PERFORMANCE REQUIREMENT

IEEE 802.11g standard supports data rates up to 54Mbits/s using OFDM modulation. To achieve the desired performance upper bound, it is crucial to have a good synthesizer performance, a good circuit linearity, a close match of I and Q signals in both the modulator and the demodulator.

5.1 Synthesizer Output Phase Noise

OFDM is recognized to be the one of the best solution for wideband wireless communication systems, but it exhibits as drawbacks a large amplitude fluctuation and a high sensitivity to phase noise. There are a lot of discussion [3, 4] about the system performance degradation from VCO phase noise. To meet the performance requirement, in the transceiver that we want to design, phase noise of Phase Lock Loop (PLL) must be low enough to meet the systematic request.

5.2 Circuit Linearity

As the OFDM signal contains subcarriers very close to each other in frequency, we can imagine that each of the 52 subcarriers of the OFDM signal is a single-tone Sine-wave such that the composite waveform in the time domain will have large peaks and valleys. If the peaks of all 52 Sine-waves should line up in time, the peak voltage will be 52 times larger than that of a single Sine-wave. In this case, the peak-to-average ratio will be 17 dB. Therefore, the transceiver must be able to accommodate signals whose peak amplitudes are 17 dB larger than the average signal. This translates into the need for a large power backoff in the transmitter and wide dynamic range in the receiver. In practice, some signal clipping can be tolerated without significantly degrading the radio performance

5.3 Modulator/Demodulator I/Q Gain and Phase Imbalance

Since IEEE 802.11g uses 64 QAM OFDM, this high bit rate communication system tolerates very little error in the modulation. This means that conventionally negligible error sources like the imbalance between the in-phase (I) and quadrature (Q) branches of the receiver and transmitter can start limiting the performance of the system [10]. This is especially true when at the same time cheaper, smaller; more robust and more versatile analog front ends are required from the RF designers.

In practice, there are a lot of sources of imbalance in the receiver but they can all be reduced to three sources that can be used to model them all. Perhaps the best known source of imbalance is the phase deviation from the ideal 90° between the I and Q local oscillator signals. This phase deviation skews the I/Q planes and can be modeled by letting the I signal to leak into the Q branch, which can be seen in single carrier systems as a skewing of the constellation points. This phase mismatch is typically constant over the channel but varies from one channel to another, making it difficult to compensate with fixed correction schemes.

The second source of imbalance is the amplitude mismatch between the branches, which can be either constant or frequency dependent within the channel depending on the source of the error.

The third source of imbalance is much more serious because it treats the upper and lower sidebands differently and is impossible to detect with the known time-domain methods. This imbalance typically comes from the baseband filters when there is a mismatch of the cutoff frequencies between the two branches. It is also strongly frequency dependent and so it is impossible to correct it in time domain even if the errors were known.

6. TRANSCEIVER DESIGN GOAL

To meet the IEEE 802.11g Transceiver design goal, all the tight specifications, such as synthesizer phase noise, circuit linearity, cascade noise figure and I/Q balance, must be satisfied. We will discuss the system link-budget under systematic linearity and noise requirement in Chapter 2. As to the imbalance of modulator/demodulator on I/Q signal path, we will discuss some existing solutions on the remaining of this chapter first, to show the advantage of solution that we propose. Discussion on architecture and design of our proposed solution will be detailed in other chapters.

6.1 Solutions on I/Q Balance

To achieve the desired performance upper bound, it is crucial to deliver a close match of I and Q signals in both the modulator and the demodulator. Solutions on I/Q gain and phase mismatch compensation can be classified into three types.

The first type uses a fully-digital compensation technique. For example, Eberle *et al.* [5] presented digital compensation architecture for the I/Q mismatch problem that occurred in a digital receiver.

The second type uses baseband plus RF front-end to calibrate the gain and phase mismatch. For example, baseband signal processing techniques have been used to generate a test signal to the RF front-end [6, 7]. Since the output of the RF front-end running this test signal will give information on the I/Q mismatch, the baseband can sense this output to determine a control code for the front-end circuits to compensate the I/Q mismatch.

But the RF front-end and baseband circuits are often developed independently and thus implementing the second type of calibration is not possible. The third type does not require any baseband support. For examples, the 802.11a/b/g transceiver presented by Ahola *et al.* [8] achieves an accuracy quadrature phase through carefully analysis and design of the I/Q generation parts of the local oscillator. Another approach of this type has been presented by Hsieh *et al.* [9]; an auto-I/Q calibrated transceiver with no more than 5% additional chip area for the calibration circuitry. This book is giving a more detailed description about Hsieh *et al*'s work, especially on the auto-calibration mechanism and calibration circuits design.

Chapter 2

TRANSCEIVER ARCHITECTURE DESIGN

The first section of this chapter compares the pluses and minuses of different receiver architectures. A transceiver includes two components: a receiver and a transmitter, but this chapter will only make comparisons to receiver, because the characteristic of transmitters are similar to receivers. The second section describes the reason of architecture selection. The third section gives a brief description on the selected transceiver architecture. The fourth section discusses the IF frequency consideration. The receiver link budget and design parameters will be mentioned in the fifth section of this chapter. Finally, the transmitter link budget design and design parameters will be presented in the last section.

1. RECEIVER ARCHITECTURE

In this section, three different receiver architectures will be introduced. They are compared with respect to their integratability, achievable performance, and required building block specifications. Based on these factors, we can choose an architecture that matches our needs.

1.1 Superheterodyne Receiver

The most straightforward architecture for implementing a cellular front-end receiver is the superheterodyne receiver introduced by Armstrong in 1918. Its generic scheme is shown in Figure 2.1.

The broadband antenna signal is fed into a highly selective band-pass filter, which is used to suppress the interferers residing outside of the application band. The LNA boosts the weakest channels above the noise floor of the first mixing stage. To bring the center frequency from ω_{RF} to ω_{IF}, the RF signal is first multiplied with a sinusoid $A\cos\omega_{LO}t$ by the mixer, where the LO frequency is set to be $\omega_{RF} - \omega_{IF}$. After this operation, a wanted signal band with carrier-frequency at ω_{RF} and an unwanted signal band with carrier-frequency at $\omega_{RF} - 2\omega_{IF}$ are both translated to ω_{IF}. This unwanted signal is called *image*. The operation is shown in Figure 2.2.

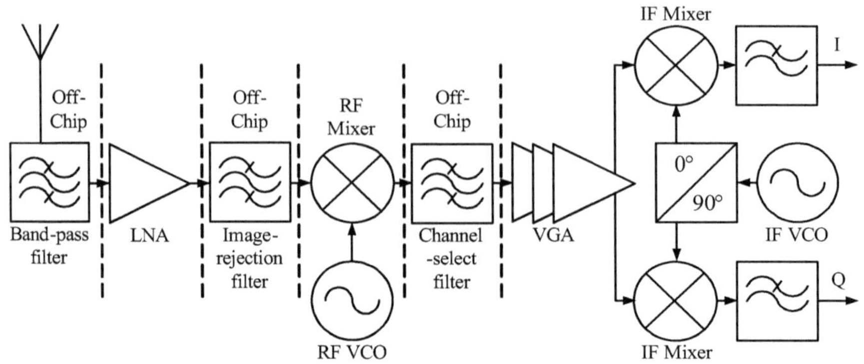

Figure 2-1. Superheterodyne Receiver Architecture.

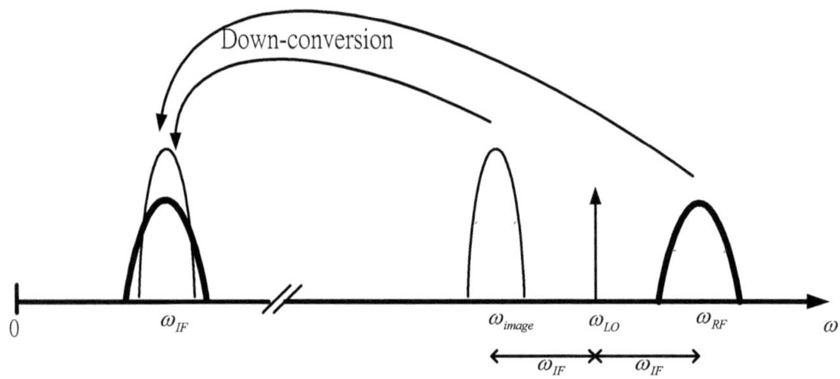

Figure 2-2. Problem Caused by Image.

The image power might be much higher than our desired signal. Before converting the signal from RF to IF, we have to put an image-rejection filter to suppress the mirror signal at $f_{image} = f_{RF} - 2f_{IF}$.

A superheterodyne receiver features a single path topology from RF to IF. Path mismatch is not an issue here, because the image rejection does not relay on any matching between the two signal paths (as in the cases of

Low-IF and Zero-IF topologies) but is done by an off-chip high-selectivity image-rejection filter or by using on-chip image-rejection technique [11]. Another important property is that the channel selection process occurs before the VGA-A/D cascade. Hence the IF VGA and the A/D converter need to handle a minimum dynamic range.

Since the image-rejection filter and channel-select filter only pass the band of interest, the specifications of the involved circuits are relaxed. Additionally, critical functions are realized by passive devices, not being implemented in a chip, the power consumption of these receivers is lower in some cases.

A superheterodyne receiver heavily relies on external high-quality-factor passive filter to perform the channel selection. Since the off-chip filters need to be driven at a low impedance level (*e.g.*, 50Ohm), they generally require a specially designed interface and high quality passive devices to perform impedance transformation from a relatively high circuit impedance to the impedance required by the filter.

In the IEEE 802.11g 64 QAM in-phase (I) and quadrature (Q) demodulator design, the demodulator is rigorous to require a good matching on I/Q [10]. Under this matching condition, the superheterodyne architecture has several advantages over other architectures: (1) The down conversion is performed in IF frequency, the devices parasitic effect is not so serious as in RF frequency. So the quadrature mismatch is smaller than the down conversion performed in RF frequency, and the quadrature mismatch is easier to calibrate in a superheterodyne architecture. (2) The receiver chain gain tuning is performed in LNA and IF variable gain control amplifier (VGA), where the gain in the demodulator is fixed. When the receiver is in gain mismatch (or imbalance) calibration, its demodulator needs to be calibrated one time only. (3) The low-pass filter in demodulator is just an aliasing filter. Compared with the channel selection band-pass filter in a Low-IF architecture and the channel selection low-pass filter in a Zero-IF architecture, this filter has smooth gain and phase response. Without the filter matching calibration, the mismatch of this filter will not cause problem in demodulator performance.

From the above discussion, it is clear that the superheterodyne receiver architecture is not a good option regarding integratability, because it does not offer a path to full integration, it needs an off-chip filter for channel selection. But this structure has the best performance over the others (Zero-IF and Low-IF); we can check the comparison result in Table 2.1.

Table 2-1. Comparison of Receiver Architectures.

	Advantages	Disadvantages
Super- Heterodyne	1. Relaxes Specs of RF, IF and Baseband circuits 2. Constant gain in I/Q demodulator 3. Fast DC offset settling time	1. External channel selection filter
Low-IF	1. Similar Advantages to Zero-IF in Integration 2. Less Severe DC Offset and 1/f Noise Problems	1. Poor image rejection 2. High order band-pass filter 3. High power consumption on ADC 4. Non-constant gain in I/Q demodulator
Zero-IF	1. Maximum integrability 2. No image rejection problem	1. 1/f Noise and DC Offset kill the performance 2. Slow DC offset settling time 3. Non-constant gain in I/Q demodulator 4. Local oscillator pulling

1.2 Low-IF Receiver

A generic scheme of the Low-IF receiver is shown in Figure 2.3. Like the superheterodyne receiver topology, the antenna signal is first passed through a band select filter, to suppress all out-of-band blocking signals and relax the required dynamic range. After amplification by the LNA, the signal is fed to two independent signal paths to generate both in-phase and quadrature components. In a Low-IF receiver, the wanted signal is mixed down to a low, none zero IF. For instance, the IF is set to half the channel bandwidth and thus the mirror signal is the adjacent channel. In Low-IF receiver, one of the most serious problems is caused by the in-band image, it will degrade the signal to noise ratio (SNR) in mixer output. Figure 2.3 shows the general image-rejection solution on a Low-IF receiver: A complex channel selection filter is placed after the in-phase and quadrature mixers, this complex filter will filter out the image noise. The image rejection performance of this complex band-pass filter highly depends on the quadrature matching between I and Q. Since I/Q paths matching becomes extremely important, the quadrature local oscillator, down-conversion mixers, on-chip filters, and VGAs need to be designed carefully.

The filter hardware overhead is a problem in Low-IF receiver design, comparing the low-pass filter in a Zero-IF receiver with the complex channel selection filter in Low-IF; for the same frequency offset attenuation, the order of band-pass filter is double of the low-pass filter.

It is rigorous to require a good matching in I/Q demodulator under the 64 QAM modulation, the quadrature phase and gain mismatch need to be calibrated. Since the Low-IF performs the down conversion in RF frequency, the demodulator gain is not fixed for different input powers, thus the demodulator gains are different and the amplitudes of gain imbalances are also different. Those characteristics need a more complicated calibration circuit design. Similar to the phase and gain calibration, the high order channel selection band-pass filter needs to be calibrated for matching [6], otherwise, the performance will be degraded by the mismatch of filter.

It is clear that the Low-IF architecture features a higher integratability. As shown in Figure 2.3, the image rejection no longer relies on a high-quality-factor filters, thus all external filters can be removed. Owing to the absence of the external filters, the number of external pins and interaction with the package parasitic will be reduced. Sensitive nodes are not brought outside, decreasing the sensitivity to interference and crosstalk. But we need to face the I/Q demodulator mismatch and poor image rejection problems.

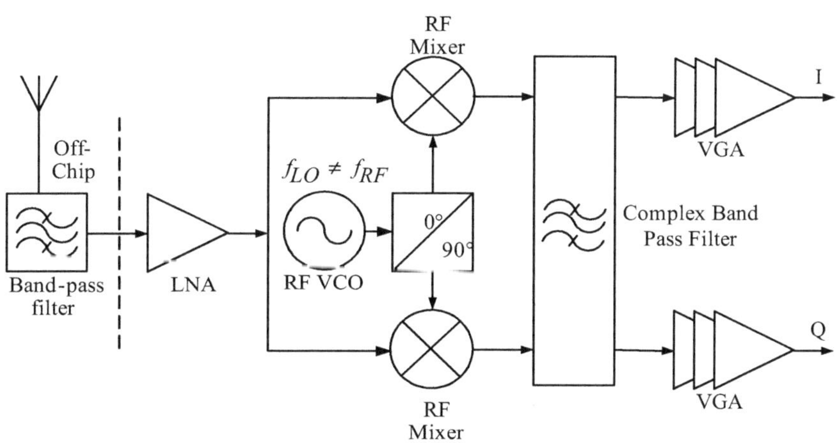

Figure 2-3. Low-IF Receiver Architecture.

1.3 Zero-IF Receiver

If the RF spectrum in a receiver is directly translated to the baseband in the first down-conversion, this type of receiver is called a "homodyne," "direct-conversion," or "Zero-IF" architecture, as shown in Figure 2.4.

The difference between a Zero-IF and a Low-IF receiver is the down-conversion step. In a Zero-IF receiver, the wanted channel is directly converted to DC, which offers two important advantages. First, the problem of image mentioned is circumvented because $\omega_{IF} = 0$. The need of image rejection in a Zero-IF system appears to be lower than that in a Low-IF system; the image power is the same as the desired signal itself, but the image of a Low-IF system is from the adjacent channel, thus the power is larger than the desired signal. Second, the IF filters and subsequent down-conversion stages are replaced with low-pass filters and baseband amplifiers which are amenable to be integrated in CMOS technology. Although Zero-IF architecture is so simple, it is not perfect in RF systems. There exist some critical issues.

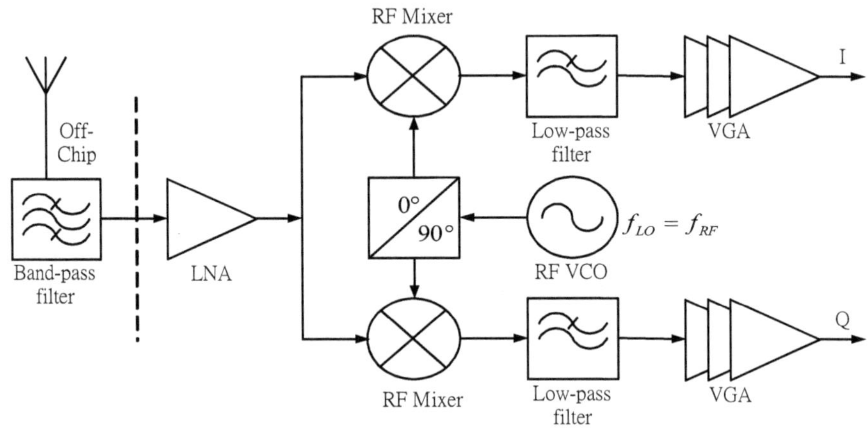

Figure 2-4. Zero-IF Receiver Architecture.

It is clear that a Zero-IF receiver is more susceptible to $1/f$ noise and DC-offset than a Low-IF receiver. The DC-offset results from the following factors: The first factor is the finite isolation between the LO port with the inputs of the mixer and the LNA. The leakage signal appearing at the input of LNA and the mixer are now mixed with the LO signal, producing a DC component. This phenomenon is called "self-mixing". The offset voltages can corrupt the signal and saturate the following stages. The feedback mechanisms from the digital part to the analog front-end can compensate for the DC-offset. However, since this feedback loop has a finite time-constant, also part of the signal contents is canceled by the loop, further degrading the signal quality.

The second factor is even-order distortion. Typical RF receivers are susceptible to only odd-order intermodulation effects. In a direct conversion architecture, even-order distortion also becomes a problem. As illustrated in Figure 2.5, two strong interferers close to the channel of interest experience nonlinearity such as $y(t)=\alpha_1 x(t)+\alpha_2 x^2(t)$ in the LNA. If $x(t)=A_1\cos\omega_1 t+A_2\cos\omega_2 t$, then $y(t)$ contains a term: $\alpha_2 A_1 A_2\cos(\omega_1-\omega_2)t$, indicating that two high-frequency interferers generate a low-frequency beat in the presence of even-order distortion.

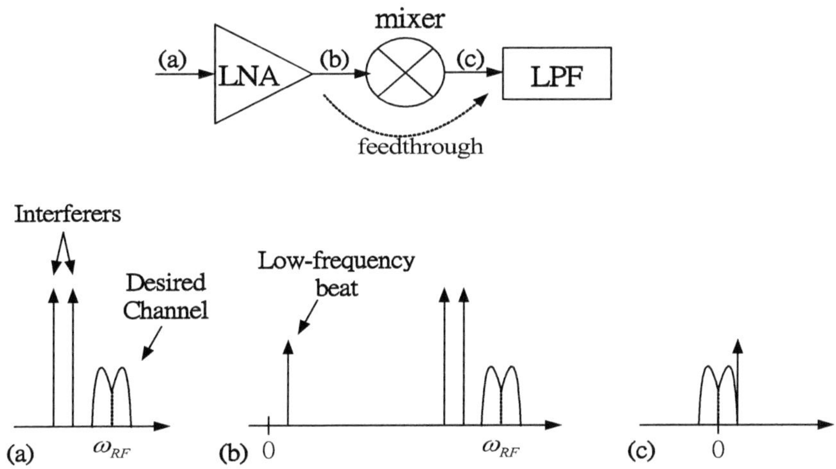

Figure 2-5. Effect of Even-Order Distortion on Interferers.

In an ideal mixer, the low-frequency beat will be translated to high frequencies and hence become unimportant after filtering. But in reality, a mixer exhibits a finite direct feedthrough from an RF port to an IF port due to element mismatch. Thus, theis mixer is not ideal anymore and produces output signal such as $v_{in}(t)(a + A\cos\omega_{LO}t)$. The low-frequency beat that appears at the output of the mixer with no frequency translation can corrupt the down-converted signal of interest.

The foregoing paragraphs will discuss the forming of low frequency interference signal on a mixer output, which is caused by the even harmonic distortion of an LNA and the finite isolation of a mixer. But a more serious problem lies in the mixer itself, if the pair of transistors in a mixer does not ideally match, the even-order intermodulation which is caused by the non-linearly performance of a mixer, will appear on the mixer output. This power will generally be bigger than the even-order intermodulation of an LNA. Table 2.1 shows the comparison summary of three different architectures.

2. COMPARSION OF OUR CHOICE

From the comparison of direct conversion and low intermediate-frequency (IF) conversion as shown in Table 2.1, we realized that the direct conversion suffers impairment of flicker noise, DC offset, even-order distortion, local-oscillator pulling and LO leakage, while Low-IF conversion is less susceptible to flicker noise and DC offset. However, Low-IF conversion does also suffer impairments of even-order distortion, LO pulling, and LO leakage. Additionally, Low-IF conversion requires stringent image rejection as an adjacent channel becomes its image, whereas direct conversion is often referred to as "no image". Furthermore, the signal bandwidth in Low-IF conversion is twice that in direct conversion, therefore requires doubling the analog-to-digital converter (ADC) sampling rate, and results in higher power consumption. Finally, the double signal bandwidth in Low-IF conversion mandates to double the baseband filter bandwidth, which further increases design complexity and power consumption.

When comparing direct conversion with the superheterodyne architecture, the latter one requires off-chip surface acoustic wave (SAW) filter and is not a preferred solution. An auto-I/Q calibration function is one of the main design goals, but when we design a direct conversion receiver, the I/Q demodulator has different gain under different input power levels. For example, if the gain range of a demodulator is 64dB with 1dB step, the demodulator will has 64 levels of gain. Under the calibration algorithm which we will mention later, the receiver needs to calibrate the demodulator I/Q gain mismatch on 64 times. It will take a long time for calibration and also need hardware to store the calibration results.

The cost of off-chip SAW filter is still a problem in superheterodyne architecture, but the cost of this filter can be compensated by the lower silicon cost obtained from a design using smaller die area and thus having higher yield rate. Moreover, System-on-chip (SoC) design is a trend under the advantages of small system size and low current consumption. Superheterodyne architecture has better noise immunity ability when compared with direct conversion, this is a very important characteristic when RF radio is integrated with noisy digital baseband processor. After we consider all the conditions as mentioned above, finally we chose the superheterodyne architecture.

3. TRANSCEIVER ARCHITECTURE

To minimize the transceiver die size, several methods have been integrated. First, the stack inductors [12] are used to replace the one-layer inductor in Low Noise Amplifier (LNA) and transmit preamplifier design; also symmetric inductors are used in VCO and VCO buffer design. Second, DC coupling instead of AC coupling is used in the inter-stage of circuits. Third, since IEEE 802.11g is a Time-Division-Duplex (TDD) system where receiver (RX) and transmitter (TX) are active on different time slots, well-designed IF and baseband circuits can be re-used in both the receiving and transmitting paths, thus reducing half of the IF and baseband circuit components.

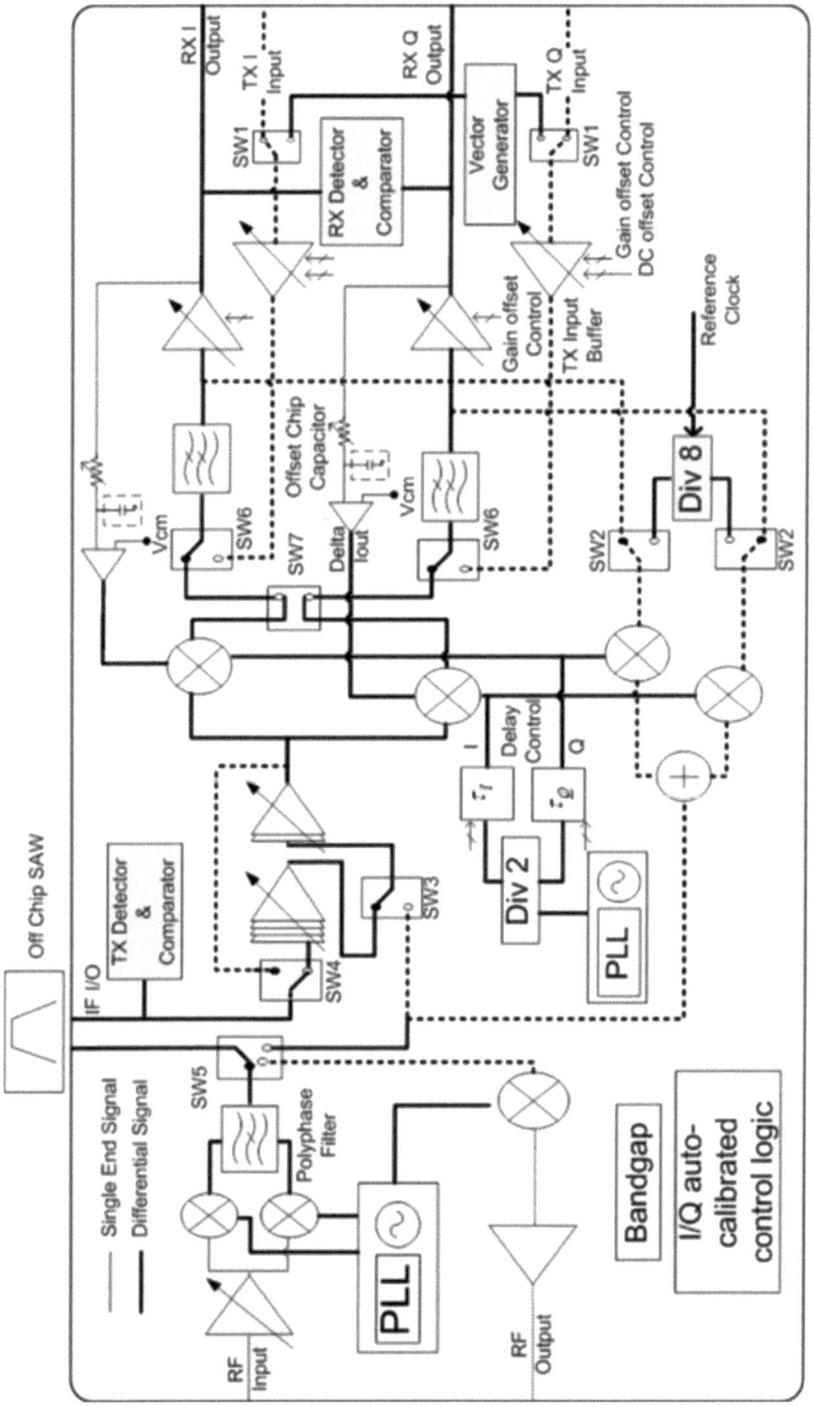

Figure 2-6. Transceiver in Receiver Mode.

Figure 2.6 demonstrates the overall transceiver architecture. Dashed lines are used to represent the transmitter signal path, and solid lines to represent the receiver signal path. The default switchers setting on Figure 2.6 is in receiver mode, at the receiver font-end, a single-end input, two-stage LNA is followed by an image-reject mixer and a polyphase filter. A CMOS analog switch (SW5) is used to drive the off-chip SAW filter. From IF to baseband, the IF signal flows through a CMOS analog switch (SW4) to a two-mode (RX/TX) Voltage-controlled Gain Amplifier (VGA), followed by an in-phase and quadrature (I/Q) down-conversion mixer and a fifth-order Bessel filter. The VGA and the Bessel filter are reused in both the receiving and transmitting paths through SW3, SW4 and SW6 to reduce the total transceiver die area. A continuous feedback DC-offset cancellation loop is designed to remove the output DC-offset voltage.

As shown in Figure 2.7, the transmitting path begins with an input buffer, followed by a RX/TX dual-mode Bessel low-pass filter (LPF). A traditional Gilbert-cell double-balance mixer, succeeding LPF through SW2, is used to up-convert the baseband signal to IF frequency. The transmitter (TX) output power range is determined by the dual-mode VGA gain that varies from -20dBm to -4dBm. The TX shares the same SAW filter with the RX using on-chip CMOS analog switches SW4 and SW5, which are designed to have a differential output impedance of 200Ω. A differential-to-single Gilbert-cell mixer drives a single-ended, two-stage preamplifier that was designed to have 21dB gain and 7dBm output P_{1dB}.

The idea of circuits reuse in both modulator and demodulator will be described in Chapter 3. The detailed circuits implementation on CMOS analog switch, two-mode variable gain control amplifier (VGA), Bessel low-pass filter and DC offset cancellation loop in demodulator will also be shown in Chapter 3.

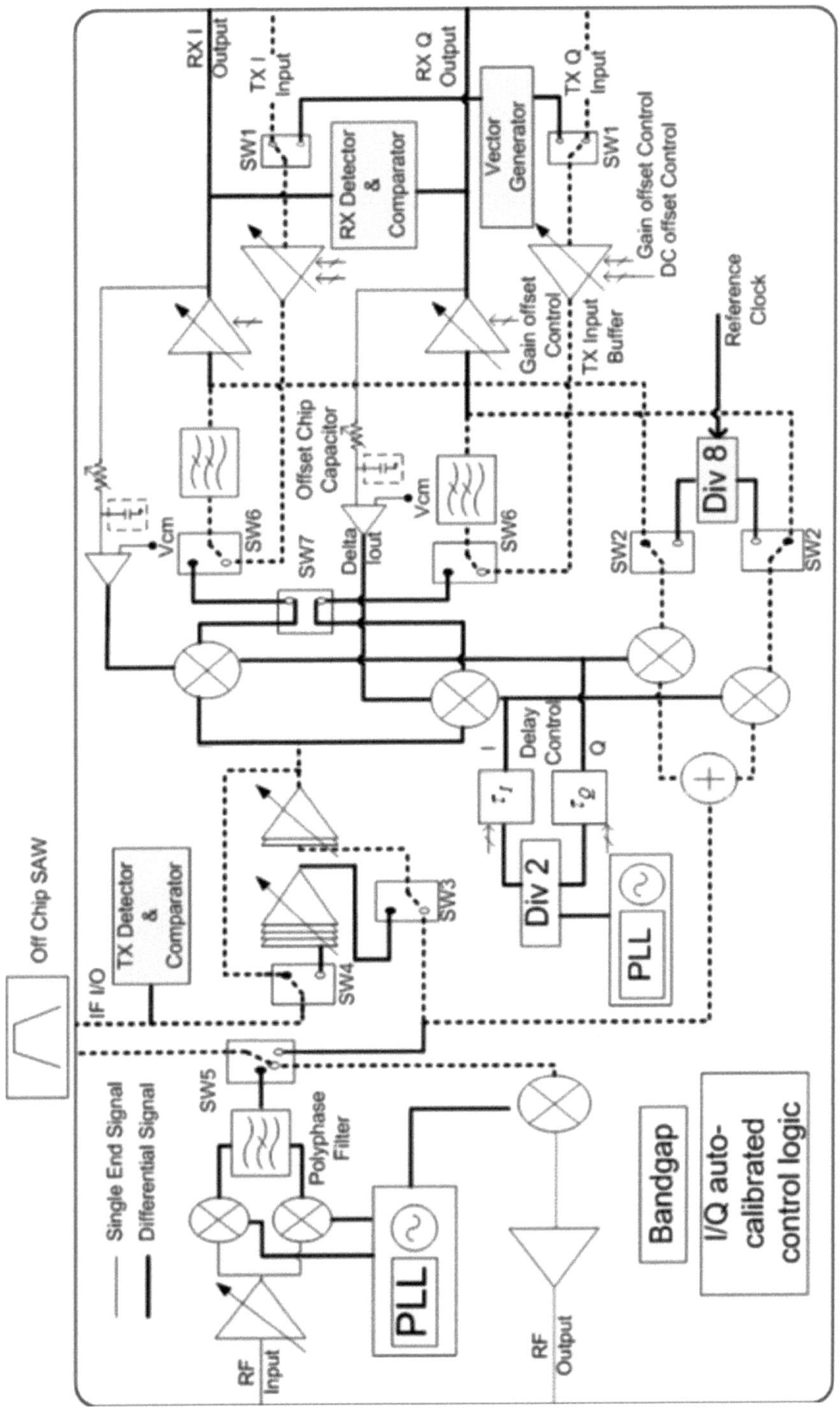

Figure 2-7. Transceiver in Transmitter Mode.

4. THE CHOICE OF IMTERMEDIARY FREQUENCY

After we chose a superheterodyne architecture for the implementation of an 802.11g transceiver, the first thing needed to decide is the intermediary frequency (IF). The trade-off in choosing the IF frequency will be shown in next paragraph.

The first thing we need to consider is the IF circuits complexity and current consumption versus the image rejection performance. If a lower IF frequency is chosen, and IF circuits can operate at this lower frequency, the complexity of circuits can be relaxed. Less current consumption is another advantage when circuits operate at a lower IF frequency. But poor image rejection performance is the problem when choosing a lower IF frequency. In our transceiver design, If IF frequency is chosen less than 340MHz, receiver performance will be degraded by the image signal, the detailed analysis will be shown later.

The channel selection filter is another consideration when choosing IF frequency. The center frequency of a channel selection filter must be the same as the desired IF frequency, and the filter bandwidth needs to meet the system requirement; when the bandwidth is too narrow, the part of desired signal will be filtered out and receiver performance will be degraded. When the bandwidth is too wide, the wide noise bandwidth and the poor adjacent channel rejection performance cannot meet the system specification. The signal bandwidth of 802.11g is 16.25MHz, so the filter bandwidth must be higher than that frequency.

In addition, the most important factor on choosing IF frequency is the spurious response location. The interference signals and their harmonics will mix with the local signal and its harmonics on the mixer, when the mixer output spurious is the same as the desired IF frequency, it will degrade the signal to noise ratio (SNR) of a desired signal.

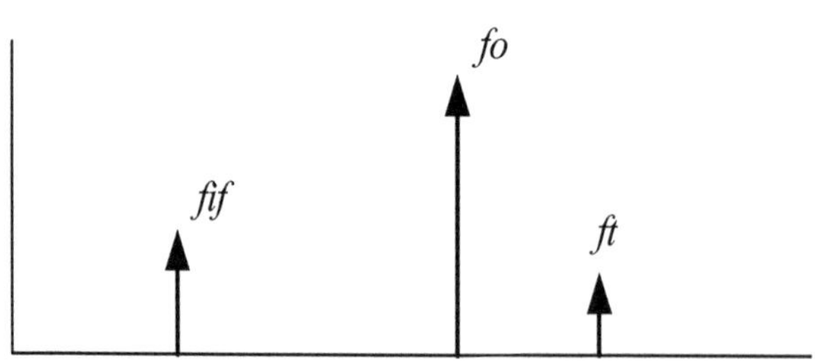

Figure 2-8. The signal relative position.

Considering the positive frequencies, *ft* = desired channel frequency, *fo* = local oscillator, *fif* = IF frequency, *fs* = the interference frequency, *O* = normalized *fo*, *S* = normalized *fs*, and *T* = normalized *ft*, where IF = 1, *S* may be solved for IF = +/-1.

In the proposed receiver architecture, the local signal is below the desired channel frequency as shown in Figure 2.8. Now we start to analyze the interference location when the interference harmonics are mixed with local harmonics and create IF spurious.

$$nfo - mfs = -/+fif \text{ (} n \text{ and } m \text{ are positive integers)} \quad (2.1)$$

$$nO - mS = -/+1 \text{ (all frequencies are normalized to } fif\text{)} \quad (2.2)$$

$$fo - ft = -fif \quad (2.3)$$

$$O - T = -1 \text{ (all frequencies are normalized to } fif\text{)} \quad (2.4)$$

By combining the two relationships of Eq. (2.2) and Eq. (2.4), eliminating *O*, an expression relating *S* to *n, m* and *T* results:

$$S = [n(T\text{-}1) +/\text{-}1]/m \quad (2.5)$$

Equation (2.5) means that when the spurious frequency is located on IF frequency, for different *n, m* and IF frequencies, what the interference frequency is. The spurious response chart is plotted in Figure 2.9. From Eq. (2.5), we sweep the parameter *T* (*ft/fo*) from 0 to 15 in X axis, for different positive integers *n* and *m* in the range from 1 to 5; we are assuming that the 6th and above harmonics powers are very small and they have no significant impact on the system. We will get different output values of *S*, the normalized interference frequency.

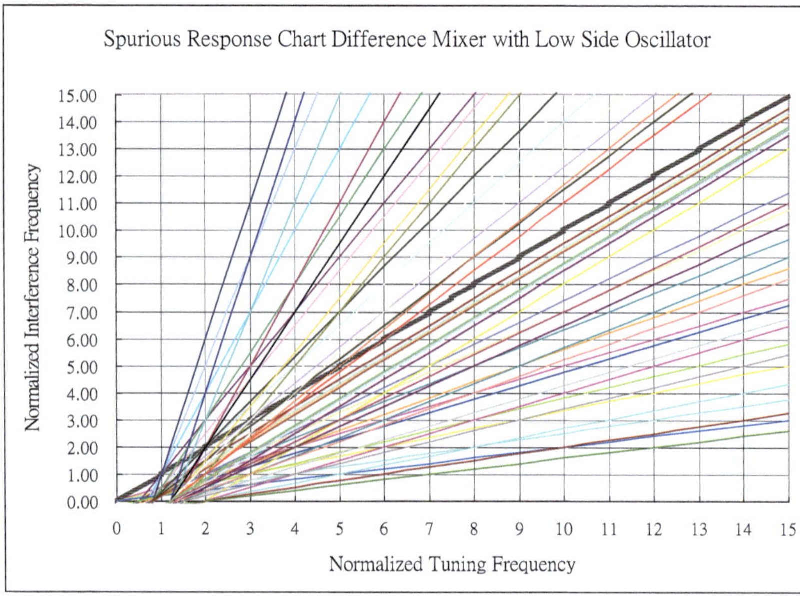

Figure 2-9. Spurious Response Chart.

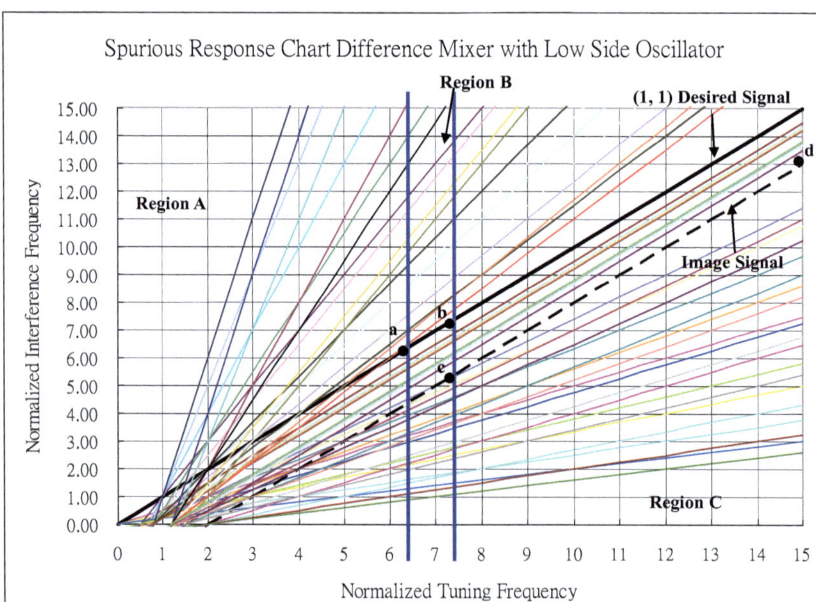

Figure 2-10. Spurious Response Chart with three Different Regions.

We can separate the normalized tuning frequency range into 3 regions as shown in Figure 2.10. In Region A, a lot of different lines cross the desired signal, it means that the spurious on IF frequency is caused by the harmonic of in-band adjacent channels mixed with the harmonic of local. Before the first down-conversion mixer, the system has a pre-select filter on LNA input, this filter can attenuate the interference signals more than 30dBc; according to the frequency offset of the interference signal. But the in-band interference will not be suppressed by this filter, so the in-band interference power can be very large. Under a non-linear LNA circuit, the harmonics power of this interference signal on LNA output will be very large, so the correspond IF spurious power will be larger too. For this case, the signal SNR on a mixer output will be degraded by this spurious.

In Region C, the normalized tuning frequency range is from 7.3 (the X axis of node c) to 15 (the X axis of node d), the corresponding IF frequency range is from *ft*/7.3 to *ft*/15. The *ft* range in 802.11g standard is from 2400MHz to 2483.5MHz [2], so the IF frequency range is from

$$2483.5\text{MHz} / 7.3 = 340.2\text{MHz} \tag{2.6}$$

$$\text{to } 2400\text{MHz} / 15 = 160\text{MHz} \tag{2.7}$$

The image signal range under this IF frequency is from

$$340.2\text{MHz} \times 5.3 \text{ (the Y axis of node } c\text{)} = 1803.2\text{MHz} \tag{2.8}$$

$$\text{to } 160\text{MHz} \times 13 \text{ (the Y axis of node node } d\text{)} = 2080\text{MHz} \tag{2.9}$$

The frequency range from 1.8GHz to 2.08GHz are the frequency bands of 2G and 3G cellphone systems, the signals crowd in this band and the signals power are also very high. If the image frequency is located in this frequency range, the IF spurious power will be large, which will degrade the desired signal SNR.

According to above discussion, the only useable frequency range is in region B, the IF frequency in region B is from node a to node b, the IF frequency in node a is:

$$2483.5\text{MHz} / 6.2 = 400.5\text{MHz} \tag{2.10}$$

From Eq. (2.6) the IF frequency in node b is 340.2MHz. Now the IF frequency which can be selected without causing spurious problem is from 340.2MHz to 400.5MHz.

The last vector which we need to consider is the harmonics of system reference clock. Not only analog circuits, the transceiver also has some digital circuits. If the harmonics of digital system clock appear on IF frequency, it will also degrade the system performance. A 40MHz clock is used as our system reference clock, the 9th harmonic of this reference frequency is located on 360MHz. So when choosing the IF frequency in the range from 340.2MHz to 400.5MHz, the IF signal band cannot be located on 360MHz.

After we consider all the conditions of spurious location, image frequency location, IF filter, PLL local frequency and reference clock harmonic, finally we choose to use 374MHz as the IF frequency.

5. RECEIVER CHAIN LINK BUDGET

This section describes link budget design of the receiver. The major requirements that a receiver must meet are sensitivity, dynamic range and adjacent channel rejection.

Table 2-2. Adjacent Channel Rejection Requirement

Data Rate [Mbps]	Adjacent Channel Rejection [dB]
1, 2, 5.5, 11 (CCK)	35
6	16
9	15
12	13
18	11
24	8
36	4
48	0
54	-1

5.1 Receiver Adjacent Channel Rejection

The adjacent channel rejection requirements of the IEEE 802.11g are shown in Table 2.2 [2]. The level of filtering, phase noise, local oscillator spurs, and linearity all affect the amount of adjacent channel rejection that a receiver provides. This receiver must be designed to ensure that the adjacent channel rejection requirements of IEEE 802.11g are met. A significant part of this is the filtering resided in the SAW filter, which should have the frequency response similar to the SAWTEK 855898 [15]. With this SAW filter, the adjacent channel interference will be removed before the IF circuits.

Before the interferences are filtered out by SAW filter, the interferences are boosted by the Front-End (FE) circuits first; the large power interferences will block the FE circuits. With the existence of these blockers, the gain of FE will decrease and FE cascade noise figure will increase, so the sensitivity becomes poor. But under our link budget design as discussed later on section 2.5.4.1, the receive FE is designed such that the high input gain compression point can avoid the blocking of adjacent channel interference.

From the IEEE 802.11g adjacent channel rejection specification, the adjacent channel must be located on a frequency of 25MHz (or higher) away from the desired signal. The voltage control oscillator (VCO) phase noise for keeping signal 25MHz away must be less than -130dBc/Hz in our design. Under this condition, the noise floor which is generated by the mixing result of adjacent channel interferences from VCO phase noise is extremely small and will not degrade the adjacent channel rejection performance.

5.2 Receiver Cascade Gain

Before the discussion on the receiver cascade noise figure, the total receive chain cascade gain needs to analyze first. To calculate the maximum receiver gain needed to cover the sensitivity requirements of IEEE 802.11b/g, the input voltage of the desired signals must be calculated. From Table 2.3, for the minimum input power –94dBm at the antenna, the input voltage is 4μVrms. Figure 2.11 shows the PCB network from antenna to chip receiver input; the signals pass through switch I, band-pass filter and finally switch II. Switch I works for the antenna diversity, the band-pass filter pre-selects the in-band signal and switch II determines whether the transceiver is in RX mode or TX mode. In a worst case system, assume that the insertion loss (IL) from the antenna to the LNA is about 3dB, the signal at the LNA input needs to have a worst-case level of 2.8μVrms (-97dBm).

The RX I/Q output is designed to accept a maximum signal of 0.5Vp. Since the receiver has been designed by assuming that the DSSS/CCK signals have a 4.5dB Peak to Average Power Ratio (PAPR) while the OFDM signals have a PAPR of 10dB [6]. So the desired output CCK I/Q signal is:

$$\text{CCK output signal } (V_{rms}) = \text{Maximum linear output } (V_{rms}) - \text{Backoff} \tag{2.11}$$

$$= 0.5V_p/\sqrt{2} \text{ - 4.5dB} \tag{2.12}$$

$$= 353\,mV_{rms}/1.68 \tag{2.13}$$

$$= 210\,mV_{rms} \tag{2.14}$$

The gain that the system needs to provide is therefore 20×log(210 mV /2.8μV)=97.5dB. However, since this is a superheterodyne system, the receiver must account for losses that are incurred in the SAW filter and in the matching network to the SAW filter, which can be up to 10dB. This increases the maximum gain needed by the system to 107.5dB. Since the input signal is 2.8 μV_{rms} , the contribution to the input signal that is caused by the thermal noise floor must be considered. The thermal noise floor at the input to the LNA is −174+10×log(22e6)=-100.6dBm or 2.1 μV_{rms}. Therefore the total input voltage at the input of the LNA is $\sqrt{2.1^2+2.8^2}$ =3.5μV_{rms}. This is a decrease in gain of about 1dB. Therefore the receiver must provide a maximum gain of 106.5dB across all process variations.

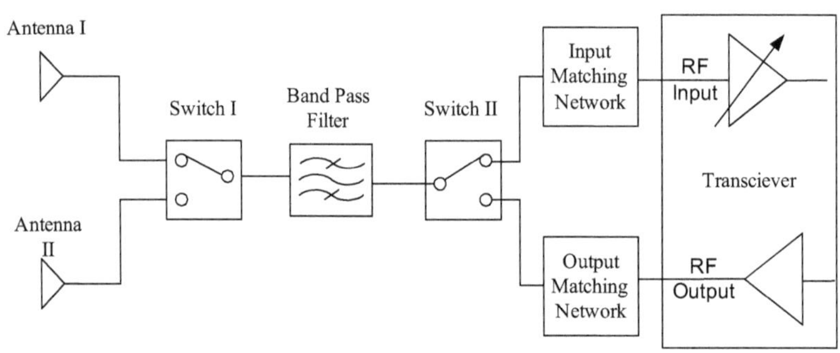

Figure 2-11. PCB Network from Antenna to Transceiver I/O.

5.3 Receiver Cascade Noise Figure

The sensitivity requirement of the receiver is shown in Table 2.3, note that this receiver is designed to exceed the sensitivity requirements as specified in the IEEE802.11g standard. How to meet the sensitivity specification? Eq. (2.15) shows the receive sensitivity calculation.

$$P_{\min} = -174dBm + 10\log BW + NF_{RX} + SNR_{\min} + IL \tag{2.15}$$

where $P_{\min}$: minimum input power (sensitivity),

-174dBm: the available noise power of the source,

BW: the bandwidth of channel selection filter,

NF_{RX} : receive cascade noise figure, and

IL: insertion loss from antenna to LNA input

Table 2-3. Receive Sensitivity Requirement

Data Rate [Mbps]	Pin at Antenna, our design requirement [dBm]	Pin at Antenna, IEEE802.11b/g requirement [dBm]
1, 2 (CCK)	-94	-80
5.5, 11 (CCK)	-84	-76
6	-87	-82
9	-86	-81
12	-84	-79
18	-82	-77
24	-79	-74
36	-75	-70
48	-71	-66
54	-70	-65

From Table 2.3, IEEE 802.11g has different transmission data rate modes. These different modes are designed under different channel environments. Under a high-quality channel condition, which is in a short distance, has no fading and no interference signals, the receiver will be working on a maximum data rate of 54Mbps. Under a poor channel quality condition, the receiver has to switch to the low data rate mode. The different data rate modes are implemented by different modulation technologies: 64 QAM OFDM, 16 QAM OFDM, 4 QAM OFDM and CCK. But the receiver architecture remains the same. To calculate the receive chain cascade noise figure requirement, we use the highest data rate mode (54Mbps) for calculation.

The insertion loss from antenna to LNA is about 3dB, the off-chip SAW filter pass-band bandwidth is 22MHz [15]. The minimum signal to noise ratio (SNRmin) requirement for baseband possessor on 8% PER is 22dB in the 54Mbps data rate mode. The sensitivity requirement from Table 2.3 is –70dBm. After substituting all the parameters into Eq. (2.15)

$$NF_{RX} = P_{\min} + 174dBm - 10\log BW - SNR_{\min} - IL$$

$$NF_{RX} = -70 + 174dBm - 10\log(22MHz) - 22 - 3dB \tag{2.16}$$

$$NF_{RX} = 5.6dB \tag{2.17}$$

From Eq. (2.17), to meet the sensitivity requirement, the receive chain cascade noise figure must be less than 5.6dBm.

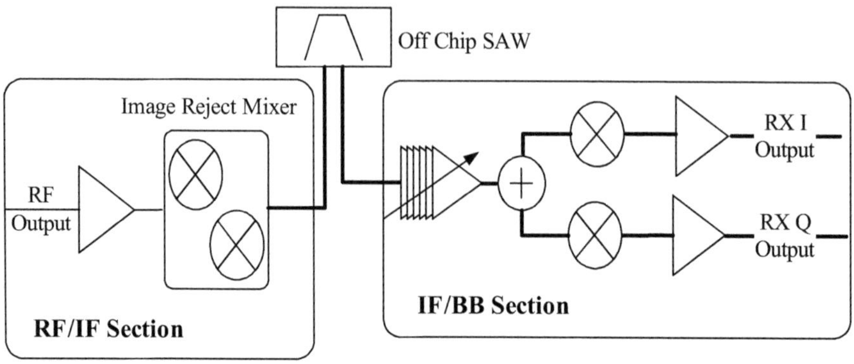

Figure 2-12. Simplified Receiver Architecture.

Now in order to analyze the receive chain cascade noise figure, we partition the receiver into two different sections as shown in Figure 2.12: the RF/IF section and the IF/BB section. The first section is the receiver Front-End (FE), which includes circuits from the LNA input to the off-chip SAW filter input. And the second section is constructed by IF VGA and baseband I/Q demodulator. From the Friis equation, the cascade noise figure of this receiver is:

$$NF_{RX} = 1 + (NF_{RF/IF} - 1) + \frac{NF_{IF/BB} - 1}{A_{P.RF/IF}} \tag{2.18}$$

where $NF_{RF/IF}$: cascade noise figure of the RF/IF section,

$NF_{IF/BB}$: cascade noise figure of the receive IF (including SAW filter), and

$A_{P.RF/IF}$: available power gain of the RF/IF section.

From the same Friis equation

$$NF_{IF/BB} = 1 + (NF_{SAW} - 1) + \frac{NF_{if/bb} - 1}{A_{P.SAW}} \tag{2.19}$$

where NF_{SAW} : noise figure of the SAW filter,

$A_{P.SAW}$: insertion loss of the SAW filter, and

$NF_{if/bb}$: cascade noise figure of the receiver IF without considering the SAW filter.

In a worst case system, assume that the cascade IF circuits noise figure $NF_{if/bb}$ is 20dB. From Eq. (2.19), the cascade noise figure $NF_{IF/BB}$ is then 30dBm. Now substituting $NF_{IF/BB}$ into Eq. (2.18):

$$NF_{RX} = 1 + (NF_{RF/IF} - 1) + \frac{30dB - 1}{A_{P.RF/IF}} \tag{2.20}$$

$$NF_{RX} = NF_{RF/IF} + \frac{1000 - 1}{A_{P.RF/IF}} \tag{2.21}$$

From Eq. (2.21), assume that $NF_{RF/IF}$ = 4, we sweep the parameter $A_{P.RF/IF}$ in 3dB step; the resulting NF_{RX} values are listed in Table 2.4.

From the calculation results as shown in Table 2.4, the RX chain cascade noise figure NF_{RX} is dominated by the RX FE cascade noise figure $NF_{RF/IF}$. When RF/IF section gain is higher than 34dB, the noise figure contribution by the IF/BB section circuitry is less than 1dB. Under this condition, we design an RF/IF section with a noise figure $NF_{RF/IF}$ of 4dB and a power gain $A_{P.RF/IF}$ of 31dB.

Table 2-4. RX Chain Cascade Noise Figure under Different RF/IF Section Gain

$A_{P.RF/IF}$ (dB)	$NF_{RF/IF}$ (dB)	NF_{RX} (dB)
25	4	7.5
28	4	6.12
31	4	5.19
34	4	4.64
37	4	4.33
40	4	4.17
43	4	4.09

5.4 Receiver Dynamic Range

According to the IEEE 802.11g specification, the dynamic range for a data rate mode of 54Mbps at the antenna input is from -25dBm to -65dBm. Our receiver is designed to exceed the dynamic range requirements as specified in the IEEE 802.11g standard. The design goal of our receiver is from -20dBm to -70dBm. The receiver has been designed assuming that the DSSS/CCK signals have a PAPR of 4.5dB while the OFDM signal has a PAPR of 10dB. So the receiver input gain compression point (IP_{1dB}) must be higher than -13dBm: -20dBm (maximum input power) –3dB (insertion loss on switches and filter) +10dB (PAPR requirement).

To cover the large dynamic range requirement, gains in the RF/IF and IF/BB sections are designed to be controllable. We will give a detailed description of gains, noises and linearity performance requirements in the RF/IF and IF/BB sections respectively.

5.4.1 RF/IF Section Gain Windows

According to the analysis in Section 2.5.3, to meet the sensitivity requirements, the RF/IF section needs to have a 4dB noise figure and a 31dB gain. Under the same RF/IF section specification, in order to meet the dynamic range requirements, the IP_{1dB} of the RF/IF section shall be higher than -13dBm. Under -13dBm IP_{1dB} and a 31dB cascade gain, the output gain compression point (OP_{1dB}) becomes +18dBm (from -13dBm to +31dB). It is impossible to implement a CMOS receiver FE with OP_{1dB} equal to +18dBm and under 2.7V power supply voltage.

From the above analysis, only one gain mode in the RF/IF section cannot cover the fully dynamic range, so we will try to design the RF/IF section with two gain modes: a high-gain mode and a low-gain mode. The first thing needs to consider in a two-gain-mode design is when the receiver changes the RF/IF section gain. When input power increases from a very small level to a high level, backoff of the RF/IF section decreases according to the input power increase. When the RF/IF section backoff is less than 10dB, the SNR of the signal starts to be degraded by circuits non-linear distortion. To solve this problem, the only way to do is to increase the RF/IF section IP_{1dB}, and change the RF/IF section from high-gain mode into low-gain mode. But after the RF/IF section gain decreases, the cascade noise figure will increase. Now another problem comes out, does the cascade noise figure in low-gain mode meet the system requirement?

From Eq. 2.18, to calculate NF_{RX}, we shall know $NF_{RF/IF}$, $A_{P.RF/IF}$ and $NF_{IF/BB}$. First, we assume that the receive gain in low-gain mode ($A_{P.RF/IF.L}$) is 5dB, if the gain in low-gain mode is too high, it cannot meet the linearity requirement, if the gain is too low, the NF_{RX} will become very poor. Second, we need to know $NF_{IF/BB}$, which depends on the gain of the IF/BB section.

From Eq. (2.14), the receiver output voltage swing is fixed to $210\,mV_{rms}$ (-0.5dBm). The cascade gain of IF/BB section ($A_{P.RF/IF}$) depends on the input power, Eq. (2.23) shows the relationship between input threshold power ($P_{IN.TH}$) and IF/BB section cascade gain $A_{P.RF/IF}$:

$$\text{Output power} = P_{IN.TH} + A_{P.RF/IF.L} + IL_{SAW} + A_{P.IF/BB} \tag{2.23}$$

$$P_{IN.TH} = \text{-0.5dB} - \text{5dB} + \text{10dB} + A_{P.IF/BB} \tag{2.24}$$

$$A_{P.IF/BB} = P_{IN.TH} + 4.5\text{dB} \tag{2.25}$$

After we derive the IF/BB section specifications on cascade gain $A_{P.IF/BB}$ and cascade noise figure $NF_{IF/BB}$, and before calculating the RF/IF section cascade noise figure requirement in low-gain mode ($NF_{RF/IF.L}$) from Eq. (2.18), we need to know NF_{RX} first. From Eq. (2.15)

$$NF_{RX} = P_{IN.TH} + 174dBm - 10\log BW - SNR_{\min} - IL \tag{2.26}$$

$$NF_{RX} = P_{IN.TH} + 174dBm - 10\log(22MHz) - 22 - 3 \tag{2.27}$$

$$NF_{RX} = P_{IN.TH} + 75.6dBm \tag{2.28}$$

From Eq. (2.18), we substitute NF_{RX} by Eq. (2.28), now $NF_{RF/IF.L}$ becomes

$$NF_{RF/IF.L} = NF_{RX} \times A_{P.RF/IF.L} - NF_{IF/BB} + 1 \tag{2.29}$$

$$NF_{RF/IF.L} = (P_{IN.TH} + 75.6dBm) \times 5dB - NF_{IF/BB} + 1 \tag{2.30}$$

The $NF_{IF/BB}$ in Eq. (2.30) depends on the IF cascade gain $A_{P.IF/BB}$. Now if we sweep the input threshold power $P_{IN.TH}$, we can get $NF_{RF/IF.L}$, NF_{RX}, IP_{1dB}, OP_{1dB} (IP_{1dB}+31dB), $A_{P.IF/BB}$ and $NF_{IF/BB}$, all the calculation results are shown in Table 2.5.

Table 2-5. Calculation Result of Important Parameters under Different Input Power

High-gain mode			Low-gain mode			
$P_{IN.TH}$	IP_{1dB}	OP_{1dB}	$NF_{RF/IF.L}$	$A_{P.IF/BB}$	$NF_{IF/BB}$	NF_{RX}
-49.57	-41.57	-10.57	-0.7	54.07	21	26.0
-47	-39	-8	23.4	51.5	22	28.6
-44	-36	-5	29.1	48.5	23	31.6
-41	-33	-2	33.2	45.5	24	34.6
-38	-30	1	36.7	42.5	25	37.6
-35	-27	4	40.1	39.5	26	40.6

From Table 2.5, if $P_{IN.TH}$ is less than -49.57dBm, $NF_{RF/IF.L}$ becomes negative, so $P_{IN.TH}$ must be higher than -49.57dBm. If $P_{IN.TH}$ is higher than -44dBm, the RF/IF section OP_{1dB} requirement is higher than -5dBm. More power is needed when high linearity performance is requested in the design of RF/IF section circuits. Considering trade-off between circuits linearity requirement in a high-gain mode and the cascade $NF_{RF/IF.L}$ in a low-gain mode, we choose to have $P_{IN.TH}$ = -44dBm. Table 2.6 summarizes the RF/IF section, noise figure and linearity specification.

Figure 2.13 shows the input power versus I/Q output SNR. To keep the demodulator I/Q gain and phase in balance, the PLL phase noise will limit the maximum Signal to Noise Ration (SNRmax) as shown in Figure 2.13. Thus, we assume that the SNRmax is about 30dBm.

Table 2-6. Receive FE Specification in Two-Gain Windows

	High-gain Mode	Low-gain Mode
Power Gain (dB)	31	5
Noise Figure (dB)	4	29.1
IP_{1dB} (dBm)	-36	-12

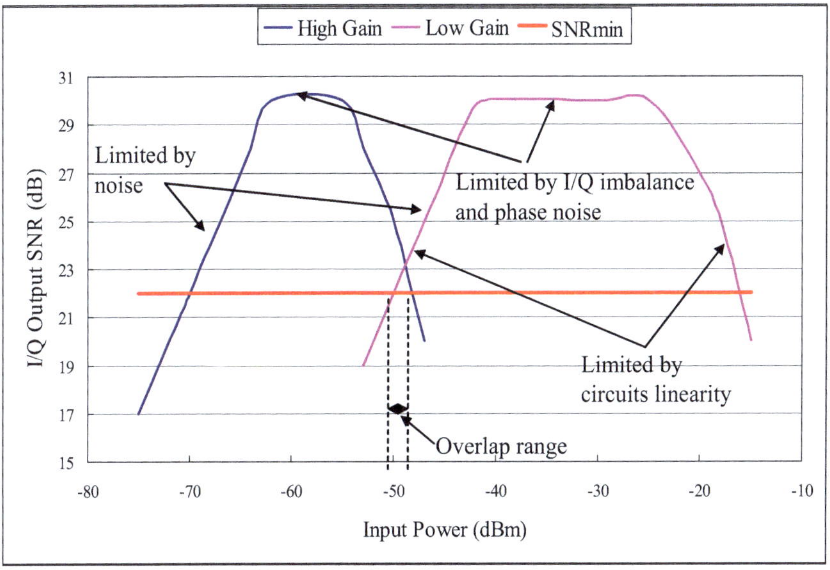

Figure 2-13. Input Power versus I/Q Output SNR when RF/IF has two Gain Modes.

When the input power is in the overlap range as shown Figure 2.13, the RF/IF section will be changed from a high-gain mode to a low-gain mode. But how does the receiver know the input power? The receiver gain is controlled by a baseband (BB) processor, which senses the demodulator I/Q output voltage swing and then determines the RF/IF section gain mode and IF VGA gain. An auto gain control (AGC) loop can thus be constructed with an RF/IF section, an IF VGA, and a baseband processor. But the problems here are the gain and noise figure of both RF/IF section and IF/BB section circuitries will vary under the process and temperature variation. For the small overlap range in Figure 2.13, there exists a dead zone where the RF/IF section is neither in high-gain mode nor low-gain mode, the SNR is less than 22dB. To solve this problem, we can improve the noise performance of IF/BB section, the noise performance of RF/IF section in low-gain mode or linearity performance in high-gain mode, but the improvement is small. In a robust design, we can have an extra gain mode to cover the dead zone. Table 2.7 shows the new RF/IF section specification after a third gain mode is added. The overlap range between each gain mode enlarges a lot when we compare with the two-gain mode in Figure 2.13. But one drawback on the system where one more gain mode is added, the AGC loop settling time must be less than about 4 to 5μs on the system requirement, but the baseband processor needs more time to determine the RF/IF section gain mode when we compare the three-gain mode with the two-gain mode. Baseband processor needs to solve this problem.

Table 2-7. Receive FE specification in Three-Gain Windows

	High-Gain Mode	Mid-Gain Mode	Low-Gain Mode
Power Gain (dB)	31	16	5
Noise Figure (dB)	4	8	23.4
IP_{1dB} (dBm)	-36	-25	-12

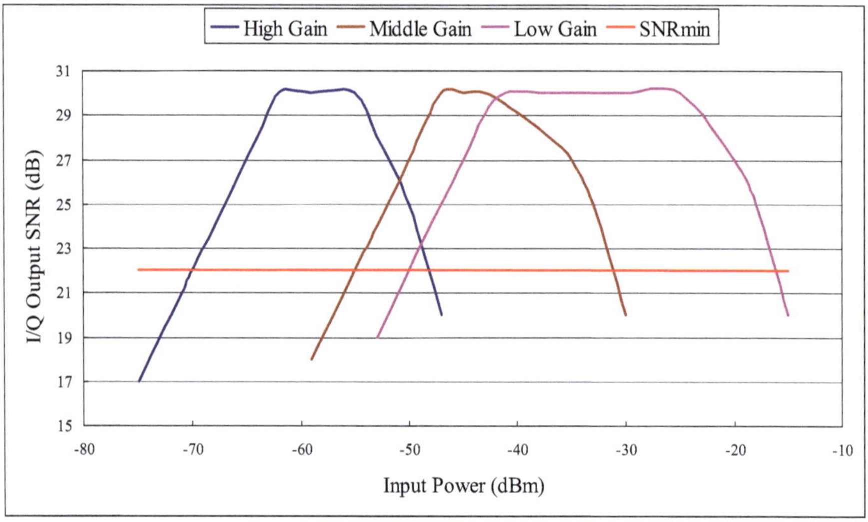

Figure 2-14. Input Power Versus I/Q Output SNR when RF/IF has Three Gain Modes.

5.4.2 Receiver IF VGA and I/Q Demodulator Specification

The most important design parameters on the IF/BB section are the linearity performance, the gain and gain range, the cascade noise figure, the low-pass filter cut off frequency, the I/Q output voltage swing, the output DC offset, the settling time of DC offset and AGC loop, the I/Q demodulator gain and phase mismatch.

The receiver is implemented in a superheterodyne architecture; the adjacent channel rejection requirement resides in the off-chip SAW filter. The aliasing low-pass filter in Figure 2.7 is used to attenuate the spurs on the mixer output, and since the filter specification is not tight, we design to have a 3dB cut off frequency at 10MHz.

For the I/Q output voltage swing and DC offset specification, the most important vector which we need to consider is the dynamic range of ADC in the baseband processor. Considering the baseband specifications, the maximum I/Q output voltage swing of a receiver is determined to be 0.5Vp and its DC offset be less than 15mV. The DC offset settling time should be less than 2μs.

5.4.3 Cascade Gain of IF/BB

From the receive chain total gain analysis in Section 2.5.2, the maximum cascade gain is 106.5dB. From the RF/IF section gain specification in Section 2.5.4.1, the maximum gain requirement in the IF/BB section is 75.5dB (106.5dB – 31dB). The maximum power at the antenna input with a 54Mbps data rate mode is -20dBm in our target, which give a -23dBm desired signal input to the receiver. The I/Q output power is fixed to -6dBm (Maximum output swing – 10dB backoff), from Eq. (2.23), the minimum gain on the IF/BB section is 19dB, so the cascade gain range varies from 19dB to 75.5dB. To compensate the process and temperate variation, the IF VGA is designed to have a 64dB gain range instead of 56.5dB; accordingly the cascade gain varies from 16dB to 80dB. The baseband processor can use a 6-bit DAC with 1dB step to control the IF VGA.

After we determinate the gain range, the next question is on the gain distribution between an IF VGA and an I/Q demodulator. If the gain of an I/Q demodulator is too high, the gain of the IF VGA becomes small. From the Frii equation, the low power gain of the IF VGA cannot suppress the noise from the I/Q demodulator. But if the gain of the I/Q demodulator is low small, the output swing of IF VGA shall be large enough to drive the I/Q demodulator; it will need a high linearity IF VGA design. Current consumption is the penalty in a high linearity circuit design. Considering the trade off between noise performance and the linearity requirement, the gain of I/Q demodulator is designed to be 13dB. So the IF VGA gain range varies from 3dB to 67dB.

5.4.4 Cascade Noise Figure of IF/BB

The receiver system is designed to ensure that the I/Q output signal SNR is higher than SNRmin in a fully dynamic range. As discussed in section 2.5.4.1, the RF/IF section has three different gain modes. With different gains and noise figures in an RF/IF section, the required gain and noise figure in the IF/BB section varies accordingly. We will investigate the noise figure requirements on different input power windows in this section.

When input power changes, the RF/IF section gain mode will change under different input power windows, the IF VGA gain will change according to Eq. (2.23), the receive chain cascade noise figure NF_{RX} will change according to Eq. (2.26). From Eq. (2.29), the corresponding IF/BB section noise figure $NF_{IF/BB}$ requirement will change too. When we sweep the input power, for the three different RF/IF section gain windows, we can get the corresponding $NF_{IF/BB}$ requirements as shown in Figure 2.15. To compare the $NF_{IF/BB}$ requirement in this figure, the $NF_{IF/BB}$ specification needs to ensure all the RF/IF section gain modes can meet the system SNRmin requirement. If the $NF_{IF/BB}$ can meet the noise figure requirement on the RF/IF section in a high-gain mode, it will also meet the noise figure requirement on the same RF/IF section in a middle- or a low-gain mode. So the "High-Gain Mode" curve will be considered as our IF/BB section noise figure specification.

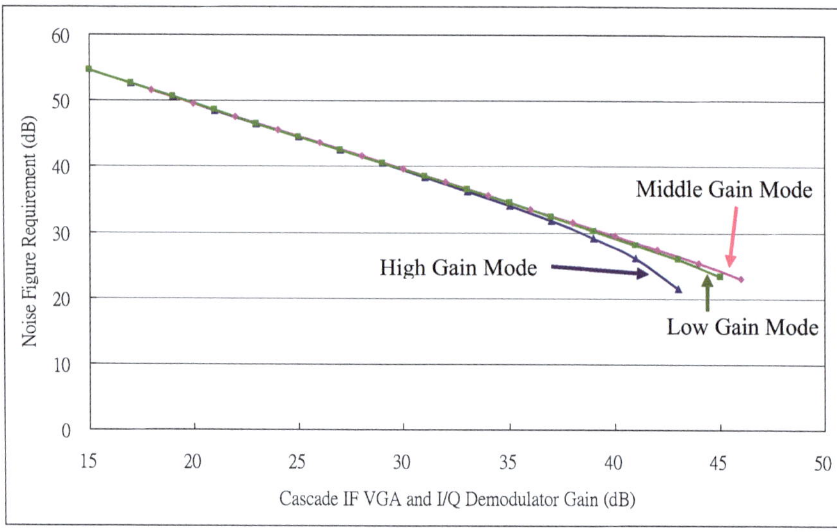

Figure 2-15. Cascade Noise Figure Requirement of IF VGA and I/Q Demodulator.

6. TRANSMITTER CHAIN LINK BUDGET

This section describes the two major requirements that a transmitter must meet. They are EVM and spectral mask, both of which drive noise, dynamic range and filter frequency response design decisions.

The following shows some of the main requirements on our transmitter. These must be met over the range of process and temperature variations that a transceiver is expected to encounter.

1. Deliver a minimum of -5dBm for all IEEE 802.11g OFDM signals and 0.5dBm for all non-OFDM mandatory IEEE 802.11b/g waveforms in all manufactured systems. This allows a system with at least 19dB of gain between the transmitter and the antenna to deliver 14dBm OFDM and 19.5dBm DSSS/CCK signals, respectively.

2. Accept differential baseband I/Q inputs of 0.5Vp.
3. Meet all IEEE 802.11b/g spectral mask requirements.
4. Meet all pertinent FCC and other regulatory bodies emission requirements. And
5. Meet all IEEE 802.11b/g EVM requirements.

6.1 Transmit Circuits Gain Distribution and Gain Range

The transmitter accepts differential I and Q signals with a maximum peak voltage (Vp) of 0.5 volt (4dBm under 50ohm). Recall from Section 2.5.4 that the DSSS/CCK signals have an *rms* signal level 4.5dB below their peak while the OFDM signals have an *rms* signal level 10dB below their peak. The transmitter is designed to keep the peak signal levels constant throughout the transmit path; independent of whether the modulation is DSSS/CCK or OFDM. Therefore, the DSSS/CCK and OFDM signals produced by the baseband processor should be scaled to take full advantage of the dynamic range of the TXI/TXQ inputs and consequently the DSSS/CCK *rms* signal level will be 5.5dB higher than the OFDM *rms* signal level.

In Figure 2.16, we partition the transmitter into two sections, the BB/IF section and IF/RF section. From the above analysis, the transmitter cascade gain requirement in a 54Mbps OFDM signals mode is shown in Eq. (2.31)

$$P_{Out} = P_{In} + 3dB + A_{P.BB/IF} + IL_{SAW} + A_{P.IF/RF} \tag{2.31}$$

$$-5dBm = -6dBm + 3dB + A_{P.BB/IF} - 10dB + A_{P.IF/RF} \tag{2.32}$$

$$A_{P.BB/IF} + A_{P.IF/RF} = 8dB \tag{2.33}$$

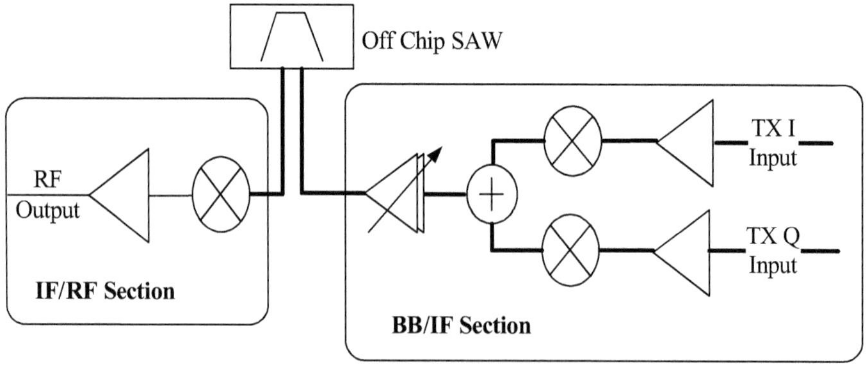

Figure 2-16. Simplified Transmitter Architecture.

From Eq. (2.33), to deliver a -5dBm output power when the input OFDM signal is -6dBm (10dB backoff from a maximum peak voltage of 0.5 volt), the 3dB in Eq. (2.31) is the power sum of I and Q path signals. Now the sum of gain will equal to 8dB. But the problem is how to distribute the gains between $A_{P.BB/IF}$ and $A_{P.IF/FE}$? First, we consider the linearity performance on the BB/IF section. After the trade off between circuits linearity and current consumption, we design to let the $OP_{1dB.BB/IF}$ at the BB/IF section be -5dBm. Under this condition, the maximum OFDM IF signal output is -16dBm. When the total I/Q OFDM signal input power is -3dBm and the IF signal output is -16dBm, the BB/IF section gain $A_{P.BB/IF}$ is -13dB. After we know the gain of $A_{P.BB/IF}$, we can get the gain of IF/RF section $A_{P.IF/FE}$ to be 21dB.

One goal on the specification of the transmitter chipset is to provide a constant output power at the antenna. To this end, the transmitter must contain a variable gain amplifier to correct the variability in the various sections of circuitry. To accomplish this, the BB/IF section of the transmitter contains an IF VGA circuit to provide a gain of -20dB to -4dB to allow all manufactured radios to deliver DSSS/CCK signals at 19.5dBm and OFDM signals at 14dBm to the antenna.

6.2 Transmit Error Vector Magnitude

In order to meet the transmit Error Vector Magnitude (EVM) requirement in the IEEE 802.11b/g specifications, one must realize that the contributors to the EVM in a transceiver are from multiple sources. The key contributors to the EVM in a transmitter are from the following sources: (1) PA non-linearities, (2) Phase noise, (3) I/Q imbalance (both magnitude and phase error), (4) On-channel noise, (5) Filtering, and (6) Non-PA non-linearities.

In order to meet the EVM requirements in the IEEE 802.11b/g specification, a transmitter must address all of the above contributors in its target specification.

The PA is the dominant consumer of current in the RF front end of a wireless LAN system. The requirement on the PA must be as much relaxed as possible in order to minimize the amount of current required and maximize its efficiency. Thus, the EVM budget is allocated such that its contribution to EVM is equal to all the other contributors in the system combined. For example, the EVM required for 54Mbps is 25dB. The contribution of the PA is therefore no worse than 28dB while the sum of all other sources must be no worse than 28dB for a combined EVM of 25dB.

Phase noise distorts the transmitted constellation in the angular dimension. For higher-order QAM constellations, which are used to achieve higher data rates, the constellation points furthest from the origin are most affected by this phase noise. The transmitter is designed to have a worst-case *rms* phase noise error of 1.5 degrees.

The I/Q gain and phase mismatches of the transmitter are specified to be less than 0.1dB of gain error and less than 1 degree of phase error. This ensures that the images are created at a level of -40dBc compared to the EVM requirement of –25dBc (*rms*). The detailed description on transmitter auto-I/Q calibration circuits will be shown in Chapter 4.

The only significant contributor of signal path noise to EVM is created in the I/Q modulaltor circuitry of the transmitter because the input referred noise of the IF-to-RF upconverter is required to be much (<-50dBc) below the signal spectrum to meet the off-channel power spectral mask requirements of the IEEE 802.11b/g standard. In order to limit the white noise contribution to EVM, the I/Q modulaltor output noise is set to a level of 32dB down from the signal. This compares to the IEEE 802.11g requirement of EVM at 25dBc (for 54Mbps). The circuitry after the SAW filter dominates the off-channel noise floor. In order to meet the IEEE 802.11g spectral mask requirements, the IF/RF section is designed to have a noise floor of –58dBc for CCK and –52.5dBc for OFDM.

To minimize the effects of non-linear phase filtering at baseband, Bessel filters are chosen, which have linear phase over the band of interest. These filters provide minimal adjacent channel filtering. The Bessel filters do remove any aliased signals that are presented on the transmitter by the baseband processor prior to up-conversion to IF.

6.3 Transmit Signal Spectral Mask

In order to meet the transmitted spectral mask requirements, the following contributors must be addressed:

1. Linearity of the signal path,
2. Off-channel noise (noise added after the SAW filter), and
3. Sidelobes of baseband signal.

Recall that OFDM has a 10dB PAPR and CCK has a 4.5dB PAPR, and the transmit path is designed such that the peak signal values of both modulation types are the same.

The off-channel noise specifications are set to ensure that the spectral mask requirements, –40dBc for OFDM and –50dBc for CCK for frequencies greater than or equal to f_C+22MHz, are maintained.

It is assumed that the baseband processor provides digital filtering to reduce the side-lobes. However, the transmitter must provide anti-alias filtering on the I and Q input signals. The dominant source of spectral shaping is the SAW filter. The SAW filter should have the frequency response similar to the SAWTEK 855898 [15].

Chapter 3

I/Q MODULATOR AND DEMODULATOR DESIGN

The first section of this chapter reviews the architecture of I/Q modulator and demodulator. To reduce the total die area, we will discuss how the IF VGA, modulator and demodulator circuits are re-used in the second section. The final section describes the proposed DC offset cancellation loop in demodulator.

1. I/Q MODULATOR AND DEMODULATOR ARCHITECTURE OVERVIEW

Figure 3.1 shows the architecture of IF VGA and I/Q demodulator in a receiver. From Section 2.5.4.3, a 64dB gain range is provided by the RX VGA, succeeding the RX VGA is the I/Q demodulator. Signals are down converted to the in-phase and quadrature DC signals by the two mixers with two different phase local signals having a phase difference of 90 degree. Following the mixer, we need an aliasing low-pass filter to remove the up-side band spur which occurs on the mixer output. This spur is far away (two time of IF frequency) from the DC signal, so using a low order filter is enough to remove this spur. To have a constant phase response, a third-order Bessel filter is designed in the demodulator. The baseband output drivers are placed after the low-pass filter to provide a non-distortion and high-swing output signal.

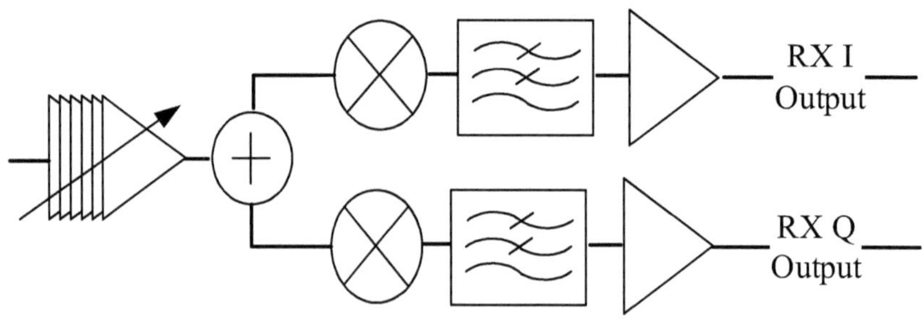

Figure 3-1. Simplified Architecture of RX VGA and I/Q Demodulator.

Figure 3.2 shows the architecture of modulator and TX VGA in a transmitter. The input buffers shift the I and Q input signal swing to a desired input signal level for the low-pass filters. The low-pass filters remove the high frequency noises which are generated by the DAC sampling clock and the noise from PCB in the transmitter input. The I and Q signals are up-converted to IF frequency by two mixers and then combined. To provide a fixed transmitter output power, we have designed a TX VGA, which follows the modulator, to compensate the variation on the insertion loss of SAW filter, variation on the off-chip PA gain, and variation of gain in the transmitting path caused by the process and temperature variation. The gain range of this TX VGA has been determined in Section 2.6.

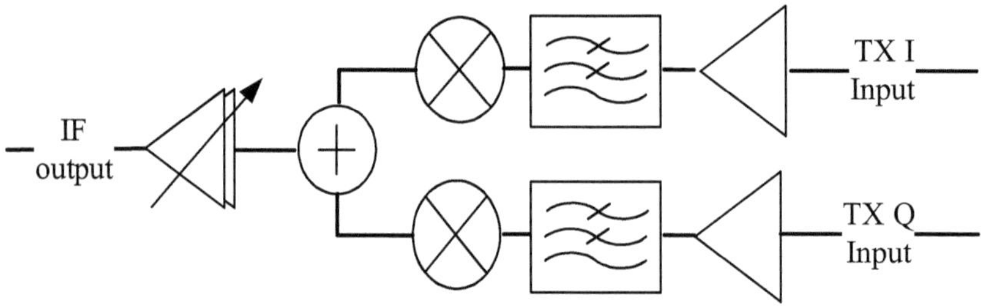

Figure 3-2. Simplified Architecture of I/Q Modulator and TX VGA.

2. VARIABLE GAIN AMPLIFIER AND LOW-PASS FILTER RE-USE

Since IEEE 802.11g is a Time Division Duplex (TDD) system where receiver (RX) and transmitter (TX) are active on different time slots, so well-designed IF VGA and baseband circuits can be re-used in both the receiving and transmitting paths, thus reducing half of the IF and baseband circuit components.

2.1 RX/TX Two-Mode Variable Gain Control Amplifier

From Figure 3.1 and Figure 3.2, we need a VGA circuit in both receiver and transmitter. The two VGA's operate in the same frequency but their gain ranges are different. In VGA design, a high gain is obtained by cascading several VGA gain cell. So the more cascade cells we use, the higher gain and larger gain range we can have. Figure 3.3(b) shows the proposed VGA cell design, the circuit is modified from Figure 3.3(a) [17]. To meet the bandwidth requirement, a source follower is added at the output of Figure 3.3(a) as shown in Figure 3.3(b). To reduce the parasitic capacitor on dominate pole (nodes A and B in Figure 3.3(b)), the common mode feedback circuits are moved from node A and node B to the source follower output in the proposed VGA cell. In our RX VGA, we cascade six VGA cells to obtain a maximum 60dB gain and a 64dB gain range, so the required gain of individual VGA cell is from -0.7 to 10dB.

From Section 2.6.1, the gain range requirement in the TX BB/IF section is from -20dB to -4dB. The total gain of the input buffer, low-pass filter and mixer, as shown in Figure 3.2, are designed to have 9dB loss. So the TX VGA gain range is -11dB to 5dB. Only two cascade VGA cells are needed to obtain a 16dB gain range.

Figure 3.4 shows the cascade RX/TX two-mode VGA design, the switches SW1 and SW2 determinate the VGA mode. AC coupling capacitors put between each of the two VGA cells are used to remove the DC offset from devices mismatch. The last two stages of VGA are used in both RX and TX. From the discussion in last paragraph, the gain range of each VGA cell in the RX mode is from -0.7 to 10dB, but the gain range in the TX mode is from -5.5dB to 2.5dB. To design a RX/TX compatible VGA cell, the gain control circuits need to be modified from [17] as shown in Figure 3.5; where the differential pair (M3, M4) and the On/Off logic control current source I2 are added to control the gain of TX VGA. When the input pin "model_ctl" is set to low, the current source I1 is turned on and I2 turned off, then the VGA is in an RX mode, otherwise the circuits are in a TX mode.

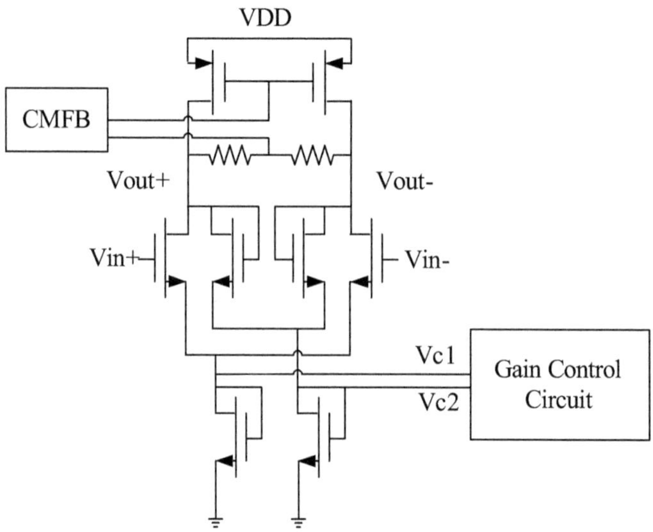

(a)

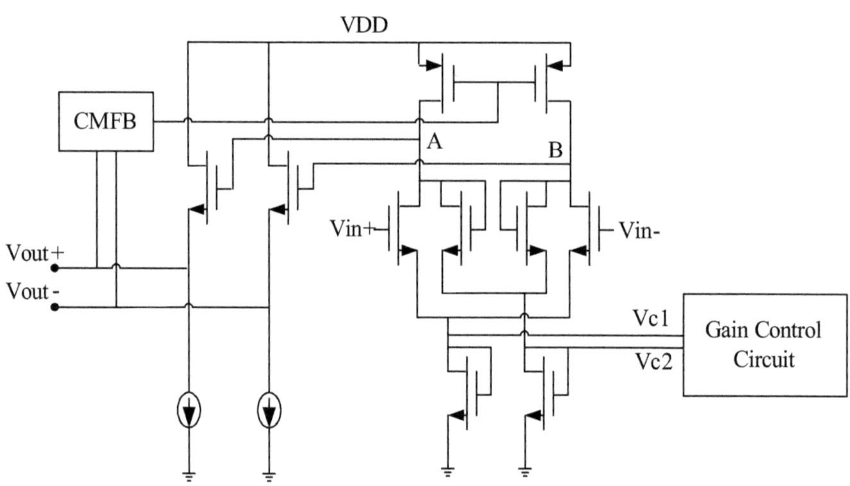

(b)

Figure 3-3. VGA Cell Design: (a) from [17] and (b) the Proposed Architecture.

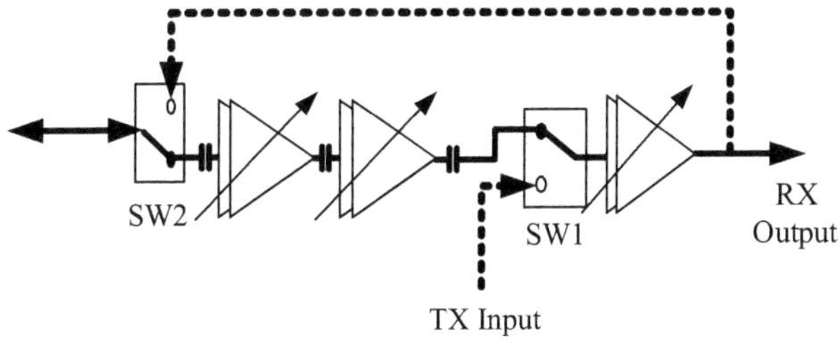

Figure 3-4. RX/TX Two-Mode Variable Gain Control Amplifier (VGA) Architecture.

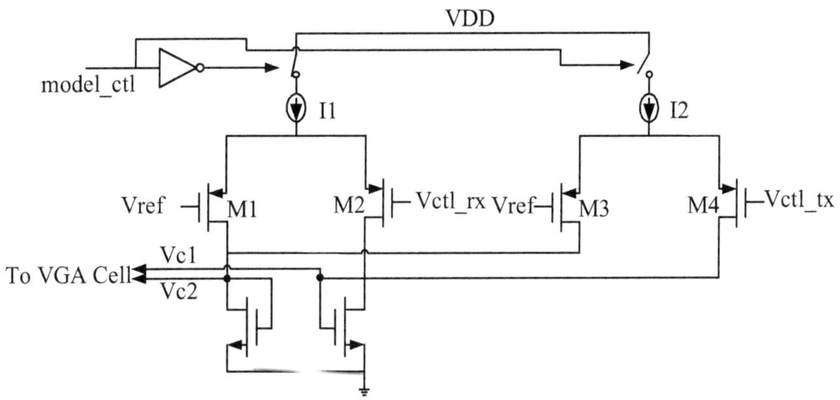

Figure 3-5. RX/TX Two-mode Gain Control Circuit.

2.2 RX/TX Two-Mode Low-Pass Filter

As discussed in Section 2.5.4.2, the RX low-pass filter is designed to remove the spurs signal on mixer output. The TX low-pass filter is designed to remove the noise from the DAC sampling clock and the PCB noise on transmitter input. A constant phase response Bessel low-pass filter can be used in both receiving path and transmitting path. Figure 3.6 shows the RX/TX two-mode third-order Bessel low-pass filter. The filter is implemented in a g_m-C structure; the OTA [18] is designed with an adaptive feedback loop to enhance the OTA cell linearity. Figure 3.7 shows the first OTA cell, the switches SW6 in Figure 2.6 are implemented by two differential pairs (M1, M2), (M3, M4) and the ON/OFF logic control current sources I1 and I2. When the mode control input "mode_ctl" is high, the low-pass filter is in an RX mode, otherwise the filter is in a TX mode.

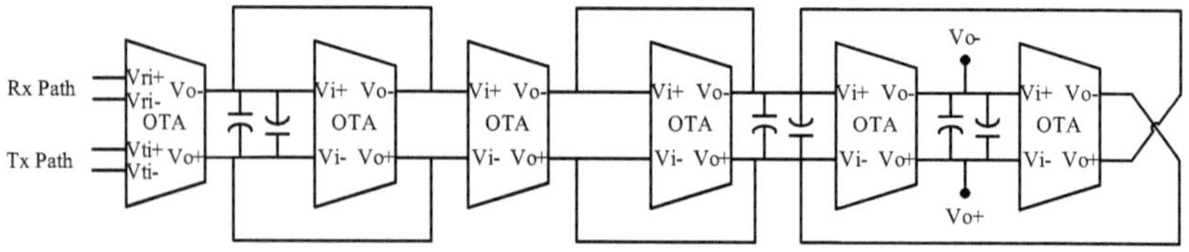

Figure 3-6. RX/TX Two-Mode Third-Order Bessel Low-Pass Filter

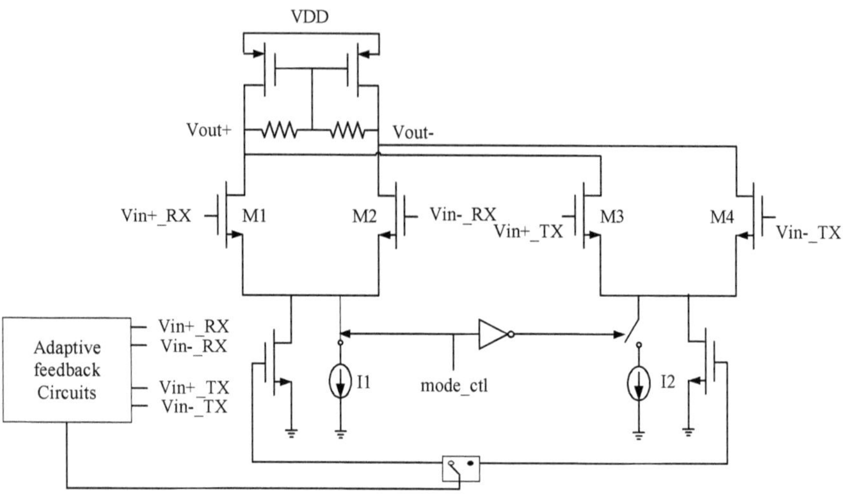

Figure 3-7. OTA Cell with two Differential Pairs for RX Input and TX Input.

2.3 DC Offset Cancellation

The finite LO-IF feedthrough in an IF mixer and the self-mixing of the LO through the IF mixers create a large DC offset voltage, the device mismatch in the circuits also creates DC offset. Those large DC offsets are possibly saturating the baseband amplifier and other subsequent stages. As depicted in Figure 3.8(a), the offset can be removed by capacitive coupling, but requiring large, relatively linear capacitors and large, low-capacitance resistors. Note that the parasitic capacitance of the resistors directly attenuates the signal. In the mixed-mode CMOS technology used here, MIM capacitor is about $1\text{fF}/\mu m^2$, translating to a very large area for the four capacitors required in the differential I and Q paths.

Another common solution on DC offset cancellation is depicted in Figure 3.8(b), the offset is reduced by negative feedback around the baseband amplifier. A critical advantage of this approach over that in Figure 3.8(a) is that parasitic capacitance of the resistors does not lower the gain at the frequencies of interest. To reduce the total die area, the grounded capacitors should be put off-chip.

The proposed DC offset cancellation architecture is shown in Figure 3.8(c), where we extract the DC from one of the differential signal output. This extracted DC voltage will be compared with the baseband amplifier output common mode voltage Vcm. Without DC offset, the extracted DC voltage is the same as Vcm.

This architecture can eliminate one loop filter compared with the negative feedback structure as depicted in Figure 3.8(b) which senses the forward path differential output for DC offset cancellation. If the off-chip loop filters are used, node A and node B in Figure 3.8(b) should connect to off-chip for the capacitive loading, but in the proposed architecture, only node A connects to off-chip as shown in Figure 3.8(c). In the differential I and Q paths, the proposed architecture can reduce two pins when compared with the conventional negative feedback structure. If on-chip loop filters are used, two large area capacitors and resistors are reduced. The proposed architecture has all the advantages as a conventional negative feedback structure. The only drawback of this structure is the double DC offset on I or Q output when compared with the conventional feedback structure.

The equivalent mixer input DC offset Δdc_{w_neg} in the negative feedback structure of Figure 3.8(b) can be obtained as:

$$\Delta dc_{w\_neg} = \frac{A}{1 + A(g_{m\beta} R_{Load\_Mixer})} \Delta dc_{wo}$$

$$\approx \frac{1}{g_{m\beta} R_{Load\_Mixer}} \Delta dc_{wo} \tag{3.1}$$

Without the DC offset cancellation loop, the equivalent mixer input DC offset voltage is Δdc_{wo}.

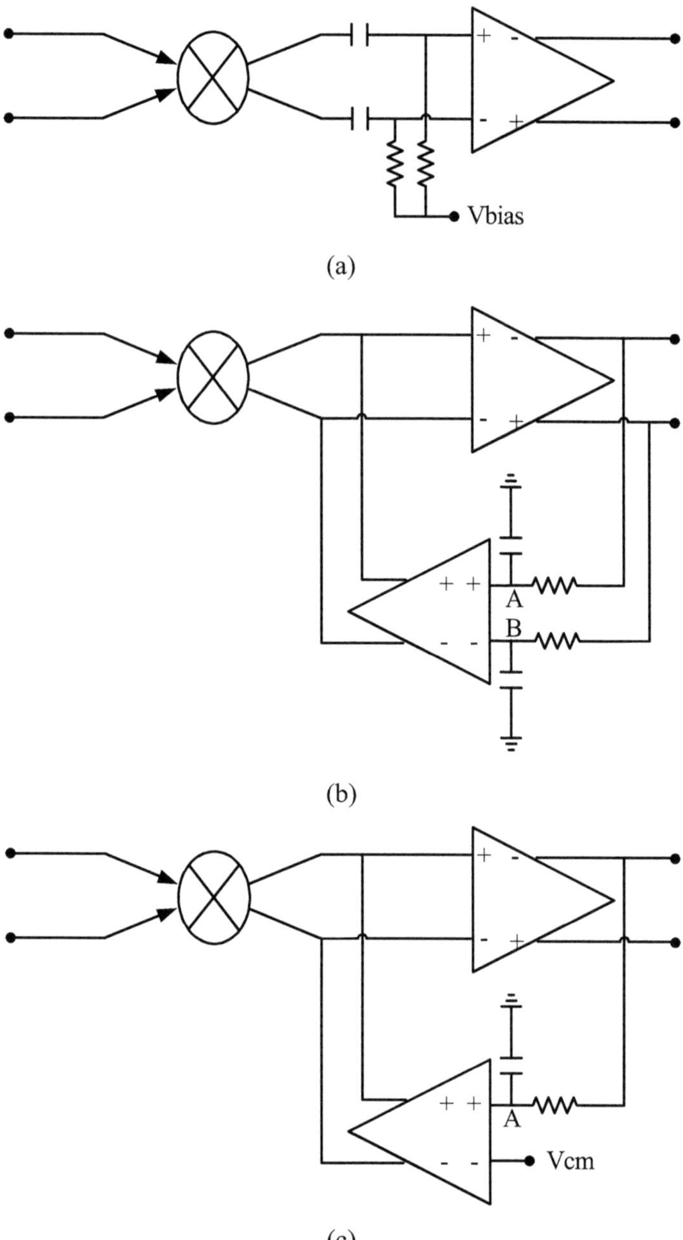

Figure 3-8. Offset Cancellation by (a) Capacitive Coupling, (b) Negative Feedback, and (c) the Proposed Single end Feedback Structure.

In our proposed structure, only one of the differential output DC is extracted, this DC will be compared with the common mode DC voltage at the differential output, and thus the equivalent mixer input DC offset is:

$$\Delta dc_{w\_our} \approx \frac{1}{\frac{1}{2} g_{m\beta} R_{Load\_Mixer}} \Delta dc_{wo} \tag{3.2}$$

The DC offset is double compared with the feedback loop gain $g_{m\beta} R_{Load\_Mixer}$, which is the penalty of this structure, but DC offset cancellation performance can be improved by increasing $g_{m\beta}$.

Since offset removal entails high-pass filtering the baseband signal, it is important to examine the consequences of such an operation for the modulation schemes of interest. In IEEE 802.11g, the center subchannel is unused, providing an empty spectrum of 156.25 kHz after translation to the baseband. Thus, if the corner frequency of the high-pass filter (HPF), falls below this value, then the spectrum of the subchannels carrying information remains intact. A corner frequency of 156 kHz requires resistor and capacitor values on the order of a few hundred kilohms and a few picofarads, respectively—relatively practical values for integration.

An important concern on high-pass filtering is the effect of frequency offsets. Since IEEE 802.11g allows a few hundred kilohertz of offset (resulting from crystal frequency inaccuracies in transmitters and receivers), the HPF may in fact suppress the main subchannel rather than the empty one. This difficulty can be resolved by automatic frequency control (AFC), i.e., by deriving an error signal from the digital baseband processor proportional to the frequency offset and applying it to the crystal oscillator that generates the reference for the frequency synthesizer.

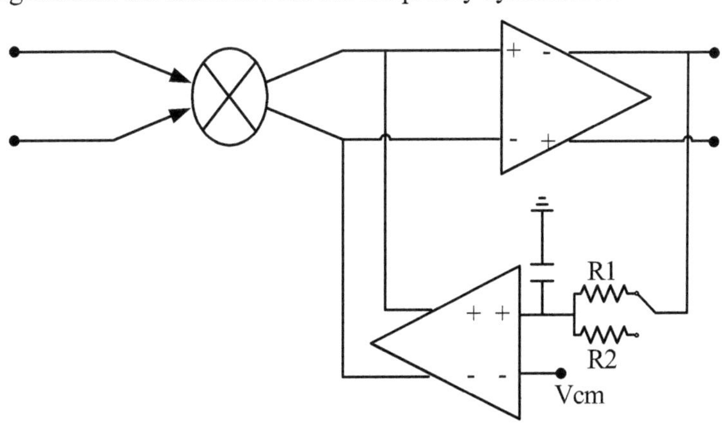

Figure 3-9. DC Offset Cancellation Loop with Two Different Loop Bandwidths.

The IEEE 802.11g standard needs to be downward compatible with 802.11b, so the demodulator needs to work with CCK QPSK modulation. A corner frequency of 156kHz will suppress the low frequency signal power when data is in a CCK coding mode. From the discussion in [19], the high-pass filter 3dB cut-off frequency is about 8.125 kHz (The data bandwidth multiplied by 0.001). An optional close-loop high-pass corner frequency is performed by switching resisters R1 and R2 as shown in Figure 3.9. It is a trade-off between non-significant signal power loss and the DC offset settling time. Whenever the transceiver changes the mode from TX to RX, the switch will connect to a small resistor R1 for fast settling (the DC is almost settling within 2μs), then a switch will change the connection back to a large resistor R2 for the 8.125 kHz HPF corner frequency requirement.

Chapter 4

AN AUTO-I/Q CALIBRATED MODULATOR

The inphase/quadraure signal processing vastly utilized in present communication transceiver faces a common issue of amplitude matching and phase matching in the inphase and the quadrature branches, and this I/Q imbalance is one of the performance bottlenecks in a transceiver. The gain and phase mismatches between inphase signal and quadrature signal degrade the signal-to-noise ratio, therefore raising the bit error rate. Thus, it is necessary to establish an auto-calibration mechanism in a transceiver, reducing the gain imbalance and phase error contributed by the I and Q paths. This chapter will describe the modulator auto-calibration mechanism first and the calibration mechanism in a demodulator will be showed in next chapter.

The first section of this chapter describes how the modulator DC offset, I/Q imbalance, and phase noise impact the system Error Vector Magnitude (EVM). The second section shows the auto-calibration mechanism and calibration circuits design. The error on calibration results which are caused by the non-ideal calibration circuits, the solution on eliminating the error will also be shown in the same section. The final section of this chapter will show the measurement result of this TX modulator auto-I/Q calibration mechanism.

1. DC OFFSET, I/Q GAIN AND PHASE IMBALANCE

Figure 3.2 shows the architecture of modulator and TX VGA, where the transmitter I and Q signals are input from the baseband processor. Two high resolution DACs are designed in baseband processor PHY to provide good matched I and Q signals for the transmitter input. The baseband I and Q signals go through the differential input buffer, low-pass filter and up-convert mixer, finally these two signals are combined on the mixer output.

As discussed in [20], the device mismatch is caused by design mismatch, layout mismatch and layout random mismatch. When device mismatch exists, the gains of I and Q paths will not be the same, the quadrature phase and power of I and Q local signals will also be different. Under the I/Q gain imbalance and quadrature phase error, the IF signal output Error Vector Amplitude (EVM) will increase.

The EVM is a measure of the difference between the reference waveform and the measured waveform. This difference is called an error vector. The EVM result is defined as the square root of the ratio of the mean error vector power to the mean reference power expressed in percentage. Figure 4.1 shows an illustration of the error vector.

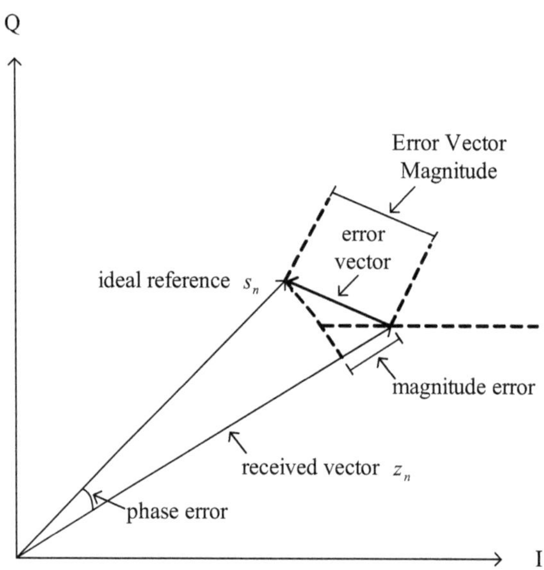

Figure 4-1. An Illustration on the Error Vector and its Components.

The measured waveform is denoted as *Z*(*k*) and consists of a signal which has been corrupted by noise and frequency offsets. After this signal has been measured at the optimal point, it will be represented as *Z*(*n*). The ideal reference signal is denoted as S_n, it is a signal free of noise and whose magnitude has been normalized to one. Z_n is referred as a modified version of the measured signal where the frequency, absolute phase, absolute amplitude and chip clock timing have been selected so as to minimize the error vector.

The instantaneous error vector is obtained by subtracting the ideal reference from the modified version of the measured waveform. The root mean square EVM is defined by

$$EVM_{RMS} = \sqrt{\frac{\sum_{n\in N} |Z_n - S_n|^2}{\sum_{n\in N} |S_n|^2}} \qquad (4.1)$$

Figure 4.2 depicts different circuit non-ideal effects on the QPSK constellation; to express the non-ideal effect, QPSK modulation is used as an example here. The ideal reference symbols are represented by a solid circle, the measured signals are represented by a small square. Figure 4.2(a) shows the DC offset effect on the constellation, all the measured symbols shift from the ideal reference symbols by an error vector E_n. Figure 4.2(b) shows the gain imbalance effect on the constellation, where the I component of an output vector is bigger than its Q component, because the gain of I path is higher than the Q path. The error vector E_n here is caused by the gain imbalance. To analyze the quadrature local signal mismatch effect on the QPSK constellation, we apply a DC test vector (V_T, V_T) to the modulator, where V_T is a DC signal. If the quadrature local signal has a phase error θ and the local frequency is w_{lo}, the modulator output symbol amplitude will be:

$$A_M = \sqrt{(V_T \times \cos(w_{lo}t))^2 + (V_T \times \sin(w_{lo}t + \theta))^2}$$

$$= V_T \sqrt{\cos(2w_{lo}) - \cos(2w_{lo} + 2\theta) + 2} \qquad (4.2)$$

When the quadrature local signal is ideally matched (θ =0), the modulator output symbol amplitude A_M becomes $\sqrt{2}V_T$. When the quadrature local signal has a phase mismatch and $\theta \neq 0$, the modulator output symbol amplitude will move away from the S_n as shown in Figure 4.2(c).

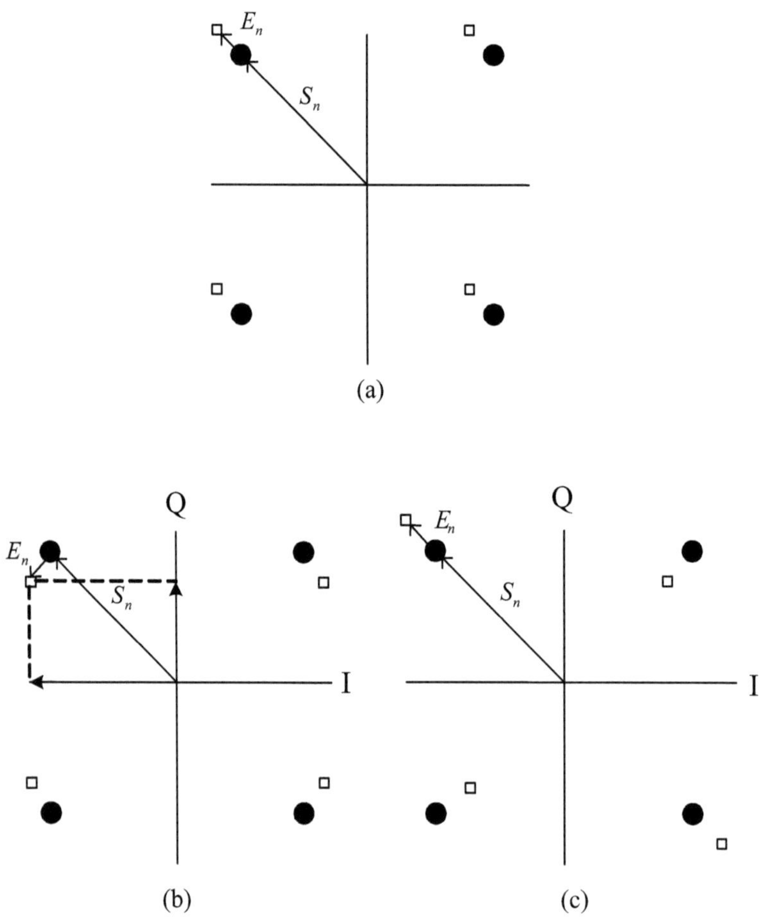

Figure 4-2. Circuits Non-ideal Effects on Constellation caused by: (a) DC Offset, (b) I/Q Gain imbalance, and (c) Quadrature Phase Mismatch.

2. DC OFFSET, I/Q GAIN AND PHASE IMBALANCE AUTO-CALIBRATION

From the discussion in [21], the calibration sequence of modulator DC offset, gain imbalance and quadrature phase mismatch must follow the flow chart which is shown in Figure 4.3. We propose to perform the demodulator I/Q calibration after the modulator I/Q calibration. The calibrations of the demodulator gain and quadrature phase mismatch are performed independently, which we will discuss in Chapter 5. Moreover, the sequence of gain and quadrature phase calibration is not important here.

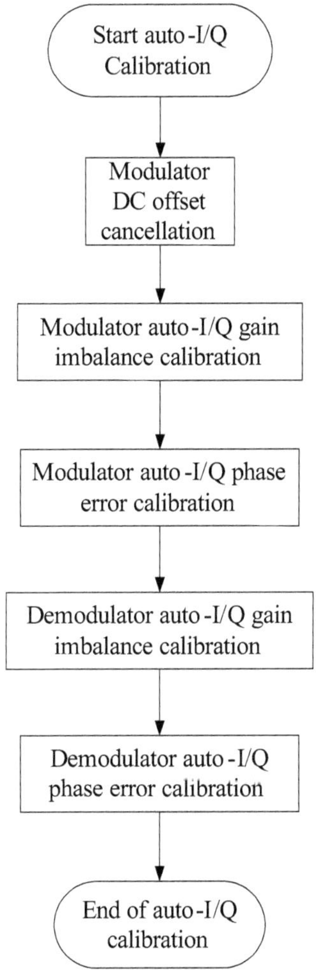

Figure 4-3. Auto-I/Q Calibration Flow Chart.

2.1 DC Offset Auto-Calibration

Figure 4.4 shows the transceiver block diagram in a modulator I/Q calibration mode. Dashed lines are used to represent the modulator calibration signal path. The active circuits are input vector generator, modulator signal path, IF VGA, TX detector and Sampling-and-Compare (S/C) comparator; all the circuits are full-filled with gray color, the other circuits are turned off. Figure 4.5 shows a simply block diagram of an embodiment of the circuits that can be used to perform modulator DC offset cancellation, I/Q gain imbalance and quadrature phase mismatch calibration. Figure 4.6 shows the flowchart of the modulator DC offset cancellation procedure.

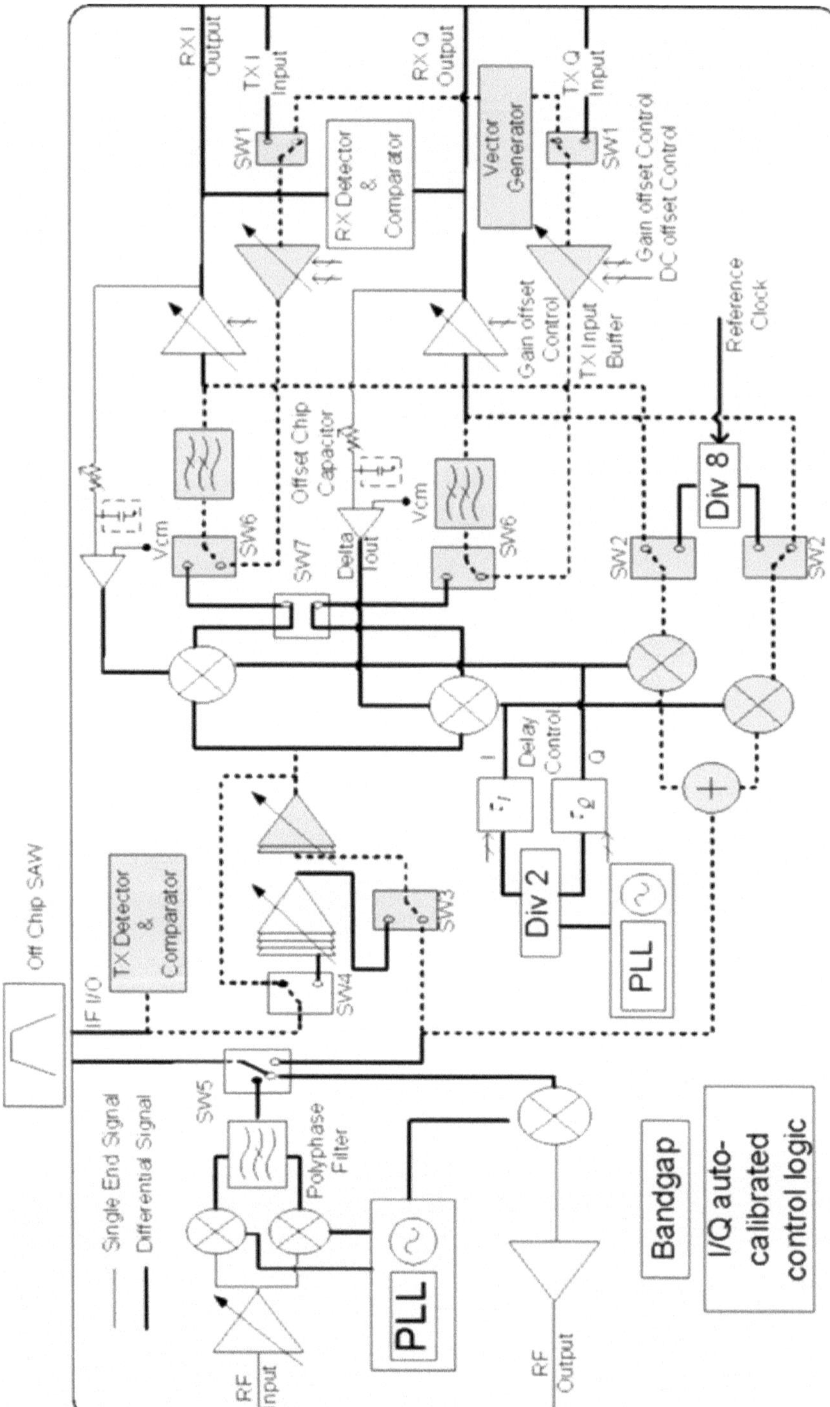

Figure 4-4. Transceiver Block Diagram in Modulator I/Q Calibration.

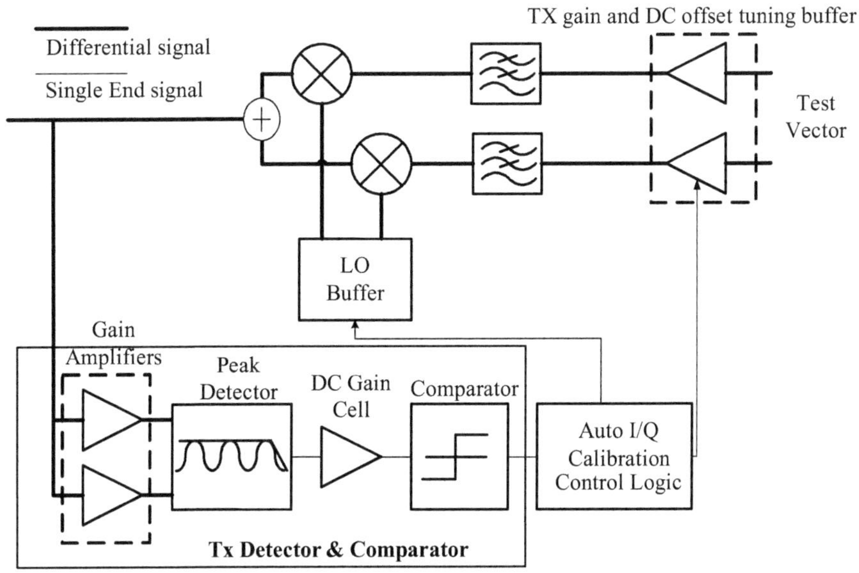

Figure 4-5. Simplified Modulator I/Q Calibration Signal Path.

The following description refers to Figure 4.5 and Figure 4.6, respectively. As shown in Figure 4.5, the circuit architecture of a TX Detector & Comparator includes two gain amplifiers, a peak detector, a DC gain cell and an S/C comparator. The DC offset cancellation should be performed before all the calibrations. As shown in Figure 4.4, a switch SW1 on both the I and Q paths switches the TX input buffer input from the baseband to the vector generator, and a DC test vector (0, 0) is applied to the modulator. A peak detector on the IF output converts the local leakage power level into a DC voltage that is then amplified by a gain cell and is applied to an S/C comparator, as shown in Figure 4.9. The auto I/Q-calibration control logic changes 1 bit in the TX input buffer DAC to create an extra DC signal on the I path. If the extra DC signal on the DAC output has the same polarity with the DC offset, the local leakage power on the peak detector input becomes bigger; otherwise, it becomes smaller. The S/C comparator gives information on the polarity of DC offset that occurred on the I path. Based on the output value of the S/C comparator, a binary search algorithm is performed to find the proper control code for DAC to create a compensated DC offset. The I/Q path DC-offset cancellation is performed in a sequence of I path → Q path → I path. A 40dBc TX local leakage can be calibrated under this algorithm.

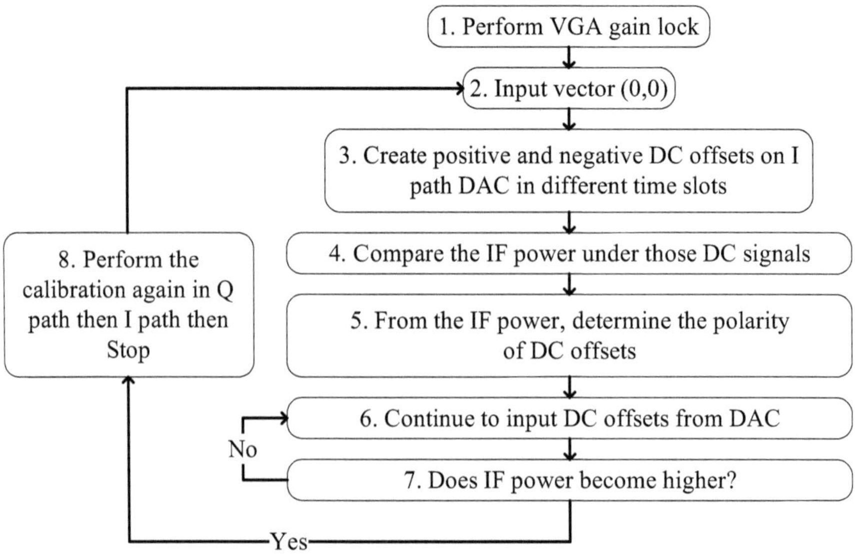

Figure 4-6. Flow Chart of Modulator Auto DC Offset Cancellation.

2.2 I/Q Gain Imbalance Auto-Calibration

Figure 4.7 shows the flowchart of a modulator I/Q gain imbalance calibration procedure. The following description refers to Figure 4.5 and Figure 4.7, respectively. As shown in Figure 4.4, a DC test vector $(0, V_T)$ or $(V_T, 0)$ is applied to the TX I/Q input in different time slots, and different signal power levels caused by the I/Q-path gain mismatch will appear on the modulator output. The peak-detector then is used to sense which test vector will give a higher output power on the modulator. And the S/C comparator output gives information on which path of the modulator has less gain. A 4-bit gain-controlled DAC is designed in the TX input buffer to compensate for the gain mismatch. The gain of the TX input buffer is then increased step by step on the smaller gain path. When the S/C comparator changes its output polarity, the I/Q path gain becomes balanced and the gain calibration process stops.

The ideal input DC test vectors $(0, V_T)$ and $(V_T, 0)$ become invalid under the existence of DC offset and vector mismatch, and thus we use equivalent test vectors $(0, V_T)$ and $(V_T + \Delta V_T, 0 + \Delta DC)$ instead. But error can be eliminated by averaging the calibration results between input vectors $(0, V_T)$ and $(V_T, 0)$; $(0, -V_T)$ and $(-V_T, 0)$. A 0.1dB gain imbalance can be achieved after the auto-calibration is performed.

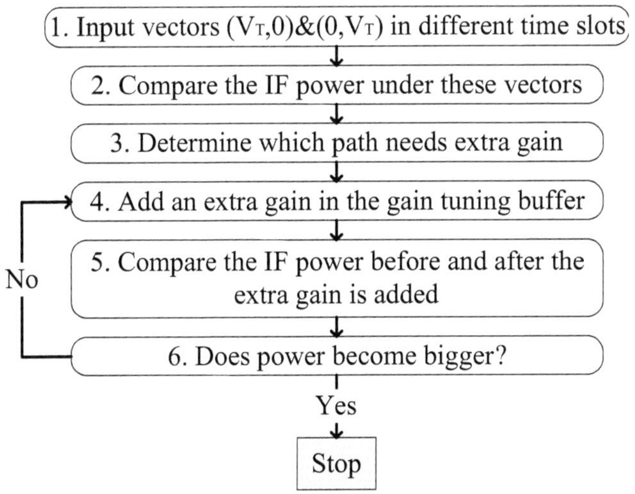

Figure 4-7. Flow Chart of Modulator Auto I/Q Gain Imbalance Calibration.

2.3 I/Q Quadrature Phase Mismatch Auto-Calibration

Figure 4.8 shows a flowchart of the modulator I/Q quadrature phase mismatch calibration procedure. The following description refers to Figure 4.5 and Figure 4.8, respectively. The quadrature phase mismatch calibration algorithm is similar to the gain imbalance calibration. As shown in Figure 4.8, the test vectors are changed to (V_T, V_T) and (V_T, -V_T) instead of (0, V_T) and (V_T, 0) in different time slots. Local quadrature mismatch causes different output power levels on the modulator output in different time slots. An extra delay is added on the local (delay-cell) buffer [22] to balance the delay. When the S/C comparator output changes its polarity, the calibration loop stops. The existing DC offset mismatch and gain imbalance on I/Q paths will generate an error in the phase calibration result, which is about 0.54° under a DC offset of 40dBc and a gain imbalance resolution of 0.1dB. The resolution of phase delay in local the buffer [22] is 0.5°, and thus the equivalent resolution of the quadrature phase mismatch is about 1.04°.

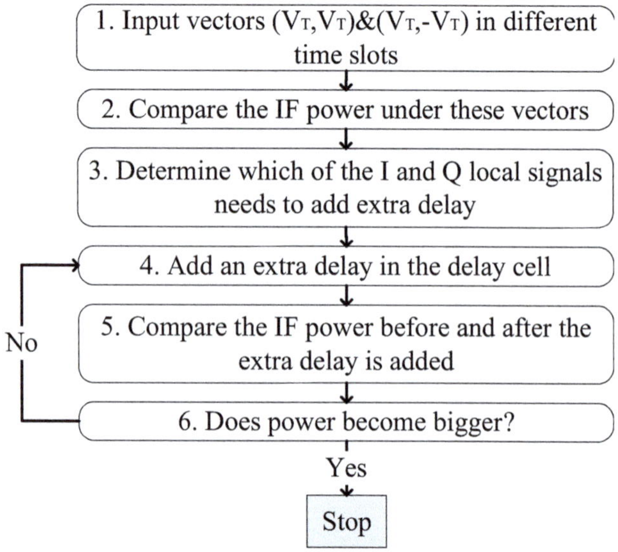

Figure 4-8. Flow Chart of Modulator Auto Quadrature Phase Mismatch Calibration.

2.4 Implementation of I/Q Auto-Calibration Circuitry

The S/C comparator circuit used in the TX calibration is shown in Figure 4.9, the S/C comparator is designed to sense the DC offset on the peak detector output under different input test vectors. To relax the resolution requirement of this S/C comparator, a fixed-gain amplifier is placed before the peak detector, and a DC gain cell amplifier is added between the peak detector and the S/C comparator. The resolution of the S/C comparator designed for I/Q gain calibration is

$$0.64mV_{(1)} \times 10_{(2)} \times 0.85_{(3)} \times 4_{(4)} = 12.8mV \qquad (4.3)$$

where

(1) is the swing of 0.1dB difference given an IF output power of -15dBm. A close loop is implemented here through the TX gain control VGA to lock the IF output power when TX is under gain and phase calibrating.

(2) is the gain of the fixed-gain amplifier between the IF output and the peak detector.

(3) is the gain of the peak detector given an IF frequency input.

(4) is the gain of the DC gain cell between the peak detector and the S/C comparator.

From Eq. (4.3), without considering the factors of fixed-gain amplifier and DC gain-cell amplifier, resolution of the S/C comparator becomes

$$0.64mV_{(1)} \times 0.85_{(3)} = 0.544mV \tag{4.4}$$

With the same architecture, the calculated resolution of the S/C comparator used in phase calibration under 0.5° resolution is 3mV. That is, our designed S/C comparator meets the sensitivity requirement of 3mV.

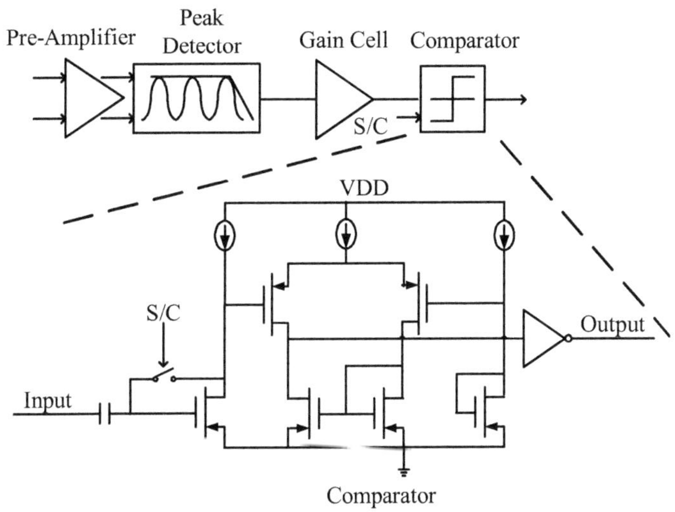

Figure 4-9. Architecture of S/C Comparator.

To compensate the quadarture local mismatch, a digital control delay cell is designed as shown in Figure 4.10. The delay cell is constructed by a forward signal path and delay signal path, the detailed circuit operation has been presented in [22]. One problem in the original delay circuit design is the output swing will change when the circuit delay is changed by the control bit. The quadrature phase mismatch calibration is done after the I/Q gain imbalance calibration, if the local power has changed, the mixer gain will also change too. To eliminate this effect, a limiter amplifier is placed after the delay cell as shown in Figure 4.10, so the output swing remains constant under different circuit delays. The delay cell is designed to have a 0.5° phase resolution.

The gain and DC offset tuning buffer (named as a TX input buffer in Figure 4.4) is shown in Figure 4.11. Different gains are obtained by the different loads of the degeneration resistors: the gain tuning buffer is designed to have a 8-level fine tuning with a 0.05dB step resolution (the seven serial resistors with RL) and a two-level rough tuning step (with or without the serial resistor Rs2). The current DAC creates a binary offset current, which flows through the resistor and creates a dc offset.

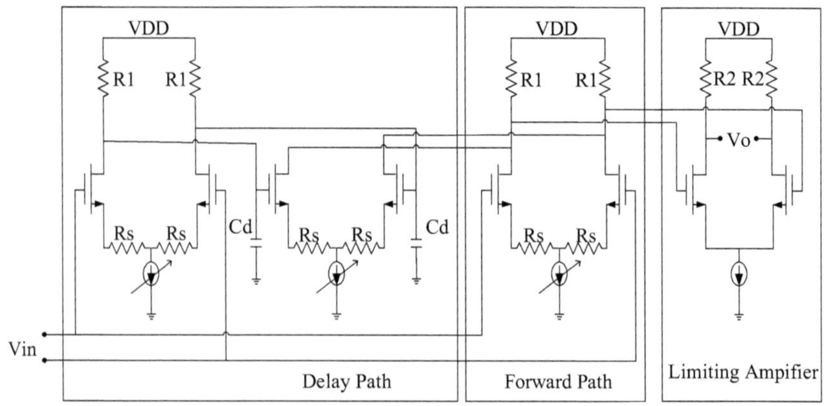

Figure 4-10. Simplified Circuit of a Delay Cell.

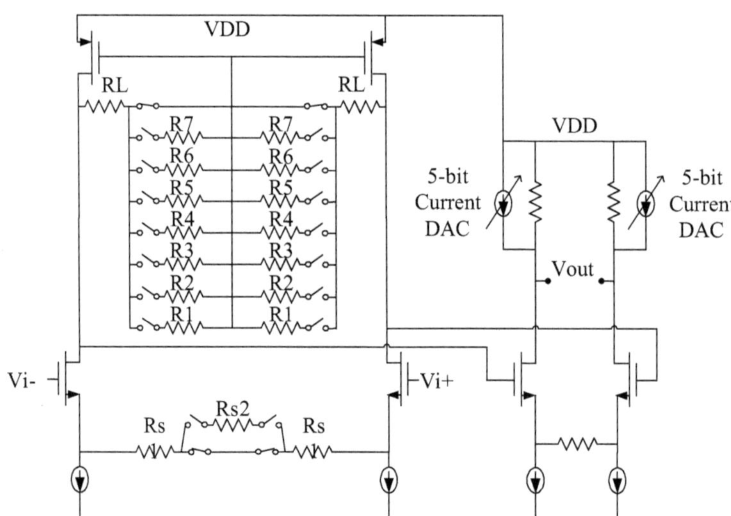

Figure 4-11. Simplified Circuit of TX input Buffer with Gain and DC Offset Tuning.

2.5 TX I/Q Auto-Calibration Measurement Result

The DC offset, I/Q gain imbalance and quadrature phase mismatch auto-calibration results will be shown in this section. Then, the modulator Single Side Band (SSB) rejection will be shown to improve the overall performance of the auto-calibration.

To track the IF output signal response when the circuits are auto-calibrating, we measure the DUT under the test setup as shown in Figure 4.12. Instead of monitoring the IF signal output, we down-convert the IF signal into a 5MHz signal. Under this frequency, it is easy to use an oscillation scope to track the signal.

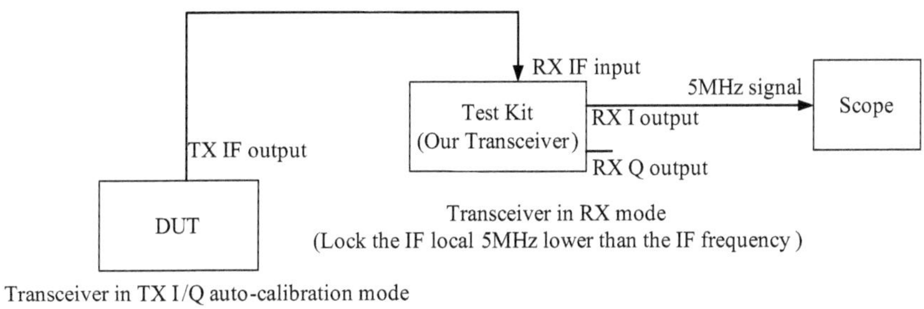

Figure 4-12. Test Environment Setup for I/Q Auto-Calibration on a TX Modulator.

Figure 4.13 shows the TX modulator dc offset auto-calibration measurement result. Before the calibration, the down-conversion output signal swing is high. During the calibration, the output signal swing becomes smaller and smaller, finally the calibration stops and output signal swing becomes very small. The small output swing here corresponds to 40dBc TX local leakage.

Figure 4.14 shows the TX modulator I/Q gain imbalance auto-calibration measurement result. As discussed in Section 4.2.2, the calibration process is divided into two segments as shown in Figure 4.14(a); the first segment of calibration is under the input vectors $(0, V_T)$ and $(V_T, 0)$, the second segment of calibration is under the input vectors $(0, -V_T)$ and $(-V_T, 0)$. Figure 4.14(b) shows a zoom-in of Figure 4.14(a), on the beginning, the detected voltage swings are different under different input vectors $(V_T, 0)$ and $(0, V_T)$. The difference in voltage swing is caused by the difference of TX modulator gains between I and Q signal paths. Following the proposed auto-calibration process, the difference in voltage swing becomes smaller and smaller, finally the voltage swings in different time slots are close to each other and the calibration stops. A 0.1dB gain imbalance can be achieved after the auto-calibration is performed. The quadrature phase calibration result is similar to Figure 4.14, but the test vector is just changed from $(0, V_T)$ and $(V_T, 0)$ to (V_T, V_T) and $(V_T, -V_T)$.

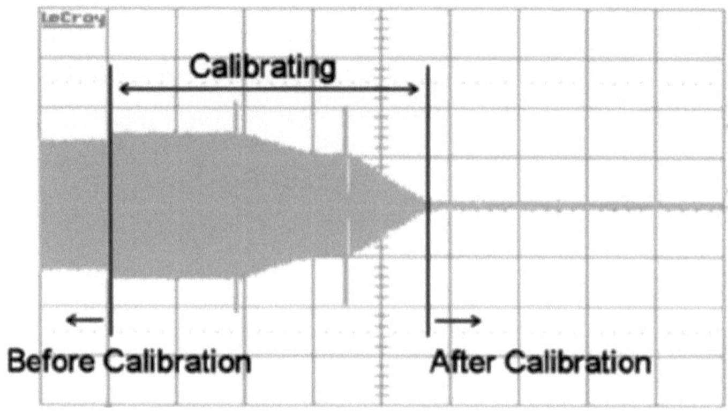

Figure 4-13. Measurement Result of TX DC Offset Auto-Calibration.

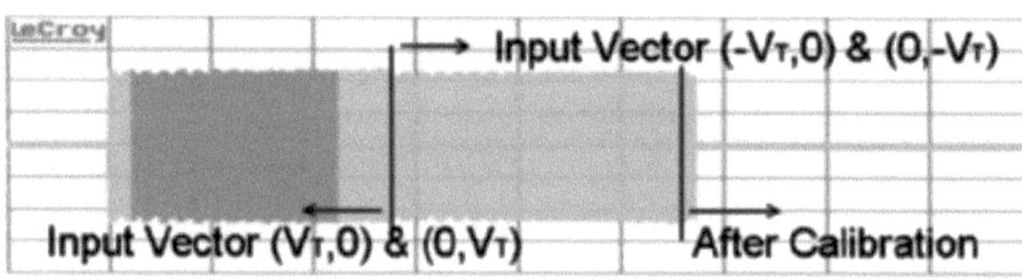

(a)

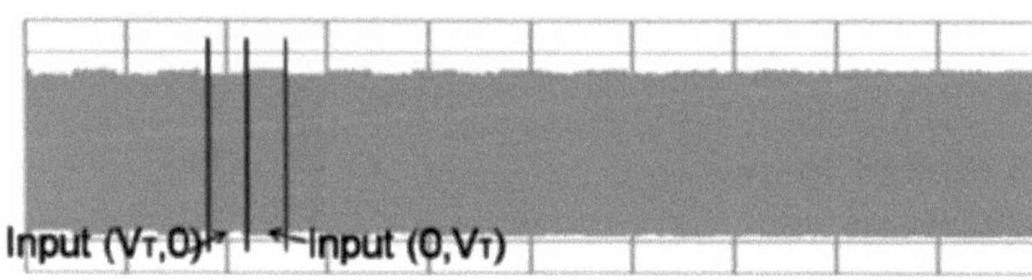

(b)

Figure 4-14. Measurement Results of TX I/Q Gain Imbalance Auto-Calibration: (a) the Whole Calibration Process and (b) the Zoom-in of (a).

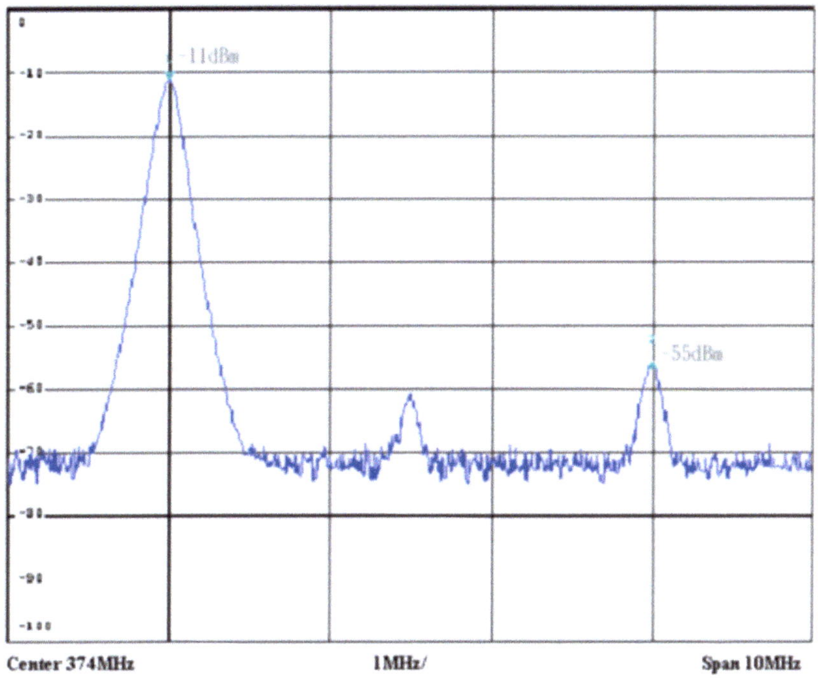

Figure 4-15. TX Modulator Single Side-Band Rejection Test.

The overall TX modulator auto-calibration performance can be measured by the Single Side-Band (SSB) rejection test. Figure 4.15 shows the SSB rejection test result, where the SSB rejection obtained is about 44dBc, as discussed in [6]. The SSB rejection performance is good enough for a 64-QAM modulation system.

Chapter 5

AN AUTO-I/Q CALIBRATED DEMODULATOR

To calibrate the demodulator, a single-tone test signal is also needed. The first section of this chapter describes how to generate this test signal in the modulator. The second section shows the auto-calibration mechanism and calibration circuits design. The error on calibration results caused by the non-ideal calibration circuits and the test signal, and the solution on eliminating the error both will be shown in this section. The final section of this chapter will show the measurement results on the RX modulator calibration.

1. SINGLE TEST TONE DESIGN

For the receiver I/Q gain imbalance and quadrature phase mismatch calibration, a single test tone is need to evaluate the receiver I/Q mismatch. From the design in [6], the baseband process will generate quadrature I and Q signals with a single frequency and these signals will be input to transmitter. Then, a loop back switch is designed in the transceiver to send this single-tone test signal to the receiver. Our proposed architecture is designed without any help from baseband processor. As shown in Figure 5.1, a single-tone signal is generated as follows. A 5MHz quadrature signal, obtained by dividing the 40MHz reference clock by 8, is applied to the up-conversion mixer through SW2. A 379MHz signal is created after up-conversion with a single-side band suppression of higher than 40dBc. Finally, this single tone will loop back to the receiver through SW5.

From Figure 5.1, the dashed lines are used to represent the demodulator calibration signal path. The calibration circuits include the single-tone signal generator, switch, RX IF VGA, demodulator signal path, RX detector and S/C comparator, these circuits are filled with gray color.

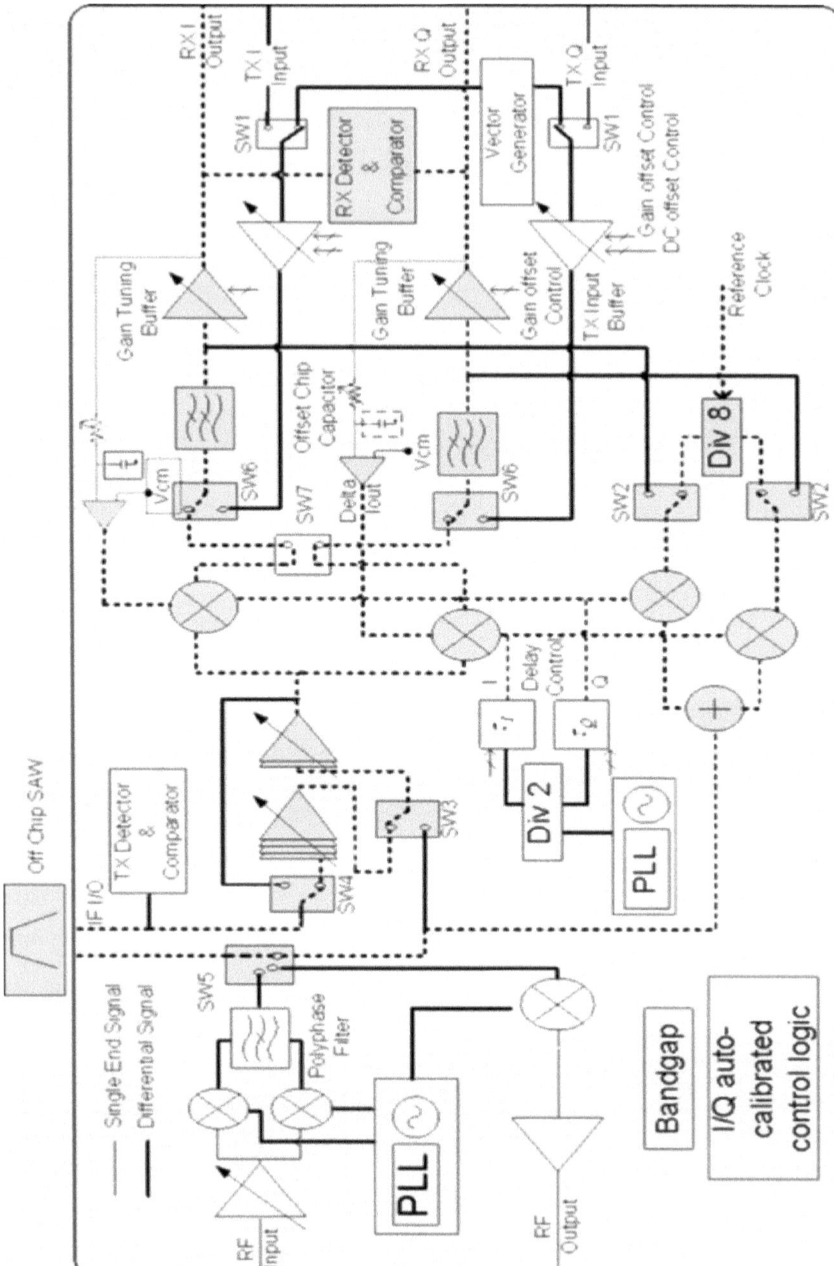

Figure 5-1. Transceiver Block Diagram in Demodulator I/Q Calibration Mode.

2. I/Q GAIN IMBALANCE AND QUADRATURE PHASE MISMATCH AUTO-CALIBRATION

As shown in the flow chart of Figure 4.3, after the TX modulator calibration, we will perform the RX demodulator I/Q gain imbalance and quadrature phase mismatch calibration.

2.1 I/Q Gain Imbalance Auto-Calibration

Figure 5.2 shows a block diagram of the embodiment of RX Detector and Comparator in Figure 5.1, and Figure 5.3 shows a flowchart of the demodulator auto-I/Q gain imbalance calibration procedure. From Figure 5.2, the Rx Detector and Comparator circuits, when in a gain calibration mode, include the first switch RSW1, a peak detector, a second switch RSW2, a gain cell, an S/C comparator, and a control logic. The mixer and gain buffer are not used in this gain imbalance calibration procedure. In demodulator gain imbalance calibration, a single-tone test signal is generated in the transmitter as discussed in Section 5.1. In the calibration procedure, the test signal goes through the normal IF/BB section, the circuits RX Detector & Comparator on the I/Q output will sense and compare the down-conversion test signal. As shown in Figure 5.2, the first switch RSW1 is switched to connect the RX I differential output to the input of peak detector and the second switch RSW2 turns to the peak detector output to make gain calibration loop. The peak detector converts the I differential signal to a DC level, being held by S/C comparator; and the S/C comparator design is the same as Figure 4.9. Then, the first switch RSW1 turns to the RX Q differential output. The Q path signal is also converted to a DC level by the peak detector and compared with the previous I signal holding by the S/C comparator to detect which path has less gain. The logic control circuit senses the polarity of the S/C comparator output signal, and generates a set of appropriate control codes to the gain tuning buffer on the path having less gain to achieve gain imbalance compensation. A sequential search algorithm is used here and calibration will stop when the S/C comparator changes its output polarity.

Figure 5.3 shows the flowchart of the demodulator auto-I/Q gain imbalance calibration procedure. Step 1 creates a test tone from TX. Step 2 performs VGA gain lock. Step 3 compares the I and Q output signal swings. Step 4 determines which path need an extra gain. Step 5 adds an extra gain in the gain tuning buffer. Step 6 compares again the signal swing. And step 7 determines whether the signal is higher than the signal in the other path; and if so, the procedure stops; otherwise, return to Step 5 and continue.

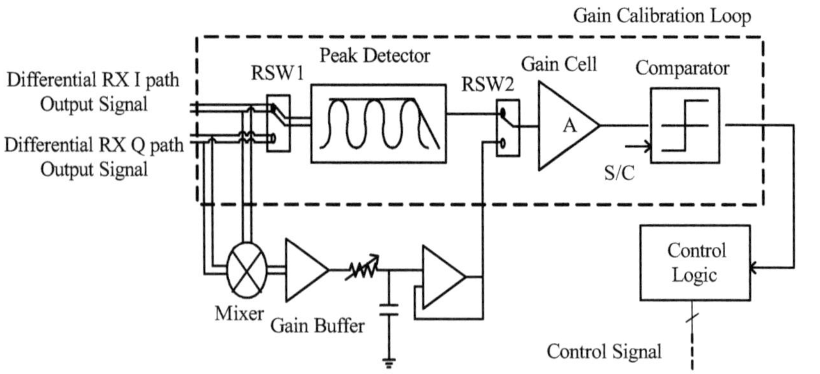

Figure 5-2. Block Diagram of RX Detector and Comparator in Gain Calibration Mode.

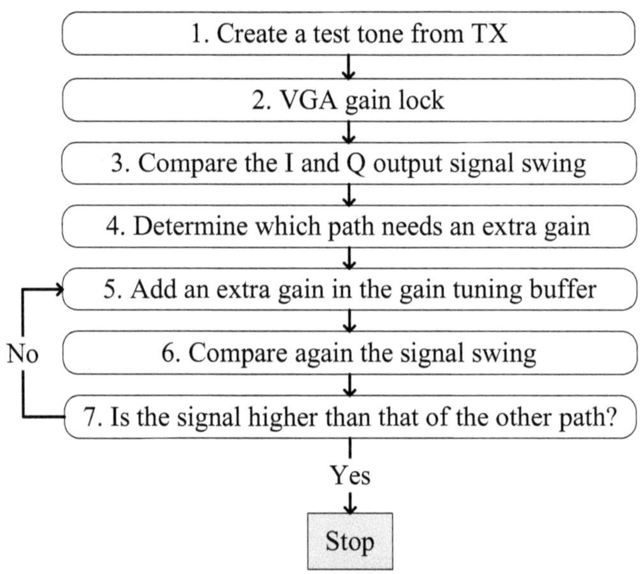

Figure 5-3. Flow Chart of Demodulator Auto I/Q Gain Imbalance Cancellation.

The non-ideal single-side band generator in the transmitter will cause error during the RX gain mismatch calibration. Assume that the ratio of the image signal to the test tone signal is *SSB*, the peak detector output is:

$$V_{Ip} \approx \frac{1}{2} \times A - \frac{1}{2} \times A \times SSB, \text{ when RSW1 switches to the I path input;} \tag{5.1}$$

$V_{Qp} \approx \frac{1}{2} + \frac{1}{2} \times SSB$, when RSW1 switches to the Q path input;

(5.2)

The compensated gain becomes:

$$\Delta G_1 = V_{Qp} - V_{Ip} = \frac{1-A}{2} + \frac{1+A}{2} \times SSB \approx \frac{1-A}{2} + SSB, \tag{5.3}$$

where A is the I/Q path gain ratio (A=1 in a balanced gain case).

The purpose of RX gain calibration is to compensate the gain imbalance caused by the term (1-A), but in this case the image signal in SSB will affect the calibration result. To eliminate this effect, gain calibration has to be performed once again with swapped I/Q path signals by setting the SW7 (Figure 5.1) switch that follows the second down-conversion mixer. The peak detector output then become

$V_{Ip} \approx \frac{1}{2} - \frac{1}{2} \times SSB$, when RSW1 switches to the I path input; (5.4)

$V_{Qp} \approx \frac{1}{2} A + \frac{1}{2} \times A \times SSB$, when RSW1 switches to the Q path input;

(5.5)

and the compensated gain is:

$$\Delta G_2 = V_{Ip} - V_{Qp} = \frac{1-A}{2} - \frac{1+A}{2} \times SSB \approx \frac{1-A}{2} - SSB \tag{5.6}$$

Combining (5.3) and (5.6), the obtained result contains only the (1-A) term, the effect of SSB is thus eliminated.

2.2 I/Q Quadrature Phase Mismatch Auto-Calibration

Assuming that the single-tone test signal $V \cdot \cos(w_{IF}t + 2\pi \cdot 5MHz \cdot t)$, which was generated on the transmitter path, is input to the receiver for I/Q gain imbalance and quadrature mismatch calibration. After this test signal is down-converted to I and Q output and the high frequency spurs have been filtered out by the aliasing low-pass filter. The I path output signal becomes $V_I \cdot \cos(2\pi \cdot 5MHz \cdot t)$ and the Q path output signal becomes $V_Q \cdot \sin(2\pi \cdot 5MHz \cdot t + \theta)$; where θ is the quadrature phase error between I and Q local signals. If we mix these two signals, output of the mixer is:

$$V_{mixer} = \frac{1}{2} \cdot V_I \cdot V_Q \cdot \sin(2 \cdot 2\pi \cdot 5MHz \cdot t + \theta) + \frac{1}{2} \cdot V_I \cdot V_Q \cdot \sin(\theta) \quad (5.7)$$

To remove the high frequency component on the mixer output, we put an RC low-pass filter succeeding the mixer; and the output signal becomes:

$$V_{lpf} \approx \frac{1}{2} \cdot V_I \cdot V_Q \cdot \sin(\theta) \quad (5.8)$$

From Eq. (5.8), the quadrature phase error θ corresponds to a DC signal V_{lpf} at the RC low-pass filter output. The proposed calibration architecture utilizes this DC signal for the RX modulator quadrature phase mismatch calibration.

Figure 5.4 shows a block diagram of the embodiment of an RX Detector and a Comparator circuits used in the quadrature phase mismatch calibration mode, and Figure 5.5 shows a flowchart of the demodulator auto-I/Q quadrature phase mismatch calibration procedure.

As shown in Figure 5.4, the circuits used in the quadrature phase mismatch mode include a down-conversion mixer with input from the RX I differential output and RX Q differential output; an RC low-pass filter; a unity gain buffer; a switch RSW2; a gain cell amplifier; an S/C comparator and a control logic. The switch RSW1 and peak detector are not used in this procedure. During the phase calibration, first the DC operation point at output of the mixer without any input is held by the S/C comparator. Then, given the input from RX I differential output and RX Q differential output, the mixer is used to mix these I and Q signals. After I/Q mixing, the I/Q phase error will generate a DC offset as shown in Eq. (5.8) at the output of the RC low-pass filter. By comparing the DC level resulted from I/Q mixing with its DC operation point held previously, the polarity of a quadrature error can be known. According to the output of the S/C comparator, auto I/Q control logic will generate a set of appropriate codes for adjusting phase. The compensation will process through the use of a delay cell on the I/Q path. A binary search algorithm is used here and calibration will stop when the S/C comparator changes its output polarity.

Figure 5.5 shows a flowchart of the aforementioned quadrature phase error calibration procedure. Step 1 creates a test tone from TX. Step 2 performs VGA gain lock. Step 3 mixes the I and Q signals. Step 4 determines the polarity of mixed I/Q signals. Step 5 adds an extra delay in the delay buffer. Step 6 compares again the mixed signal I/Q with zero. And step 7 determines whether the signal changes polarity; and if so, the procedure stops; otherwise, return to step 5 and continue.

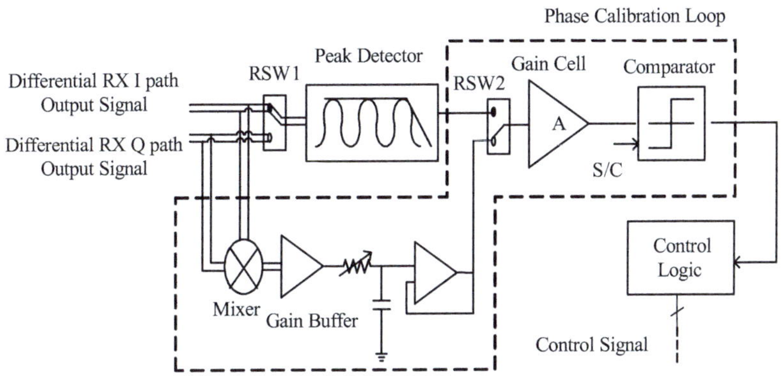

Figure 5-4. Block Diagram of RX Detector and Comparator in Phase Calibration Mode.

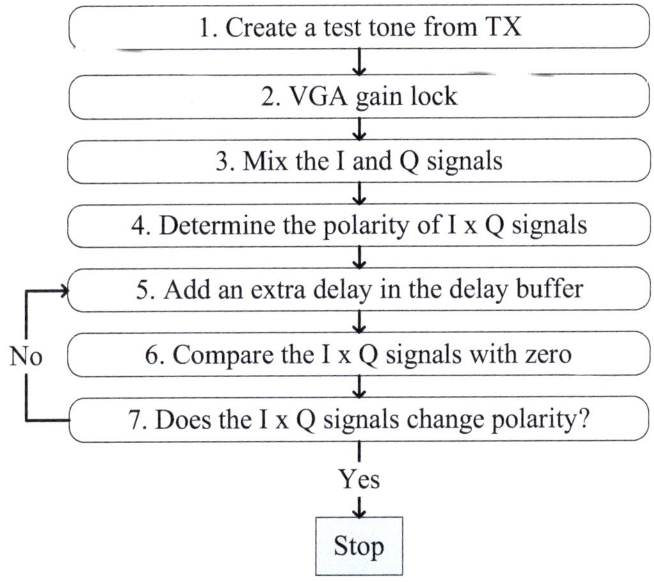

Figure 5-5. Flow Chart of Demodulator Auto I/Q Phase Mismatch Cancellation.

In this RX phase mismatch calibration case, the non-ideal gain mismatch and single-side band test tone will not affect the calibration result. The RC low-pass filter (LPF) output in Figure 5.5 is shown as follows.

Signal on LPF output $V_\theta \approx G\times\frac{1}{4}\times(A\times SSB^2 + A + A\times SSB)\times\sin(\theta)$,

$$(5.9)$$

where G=Gain of the mixer. From Eq. (5.9), since both A and SSB are coefficients of $\sin(\theta)$ only, they do not affect the calibration target, and it is thus easy to push V_θ to 0. A 1° mismatch after auto tuning can be achieved.

2.3 Implementation of I/Q Auto-Calibration Circuitry

The delay cell and gain tuning buffer design are the same as what we have shown in Figure 4.10 and Figure 4.11. The S/C comparator is the same as Figure 4.9, The resolution of the S/C comparator designed for RX gain calibration is

$$350mV_{(1)}\times 0.1dB_{(2)}\times 0.98_{(3)}\times 4_{(4)} = 16.2mV$$

where

(1) is the RX I/Q output swing. A close loop is implemented here through the RX gain control VGA to lock the RX I/Q output swing when RX is under gain and phase calibrations.

(2) is the 0.1dB resolution obtained after the gain imbalance calibration.

(3) is the gain of the peak detector given a 5MHz signal input.

(4) is the gain of the DC-gain amplifier between the peak detector and the S/C comparator.

The resolution requirement of the S/C comparator under a phase mismatch of 1° is 8mV. The resolution requirement of the S/C comparator in RX modulator calibration is more relax than the TX calibration requirement.

3. RX I/Q AUTO-CALIBRATION MEASUREMENT RESULT

Figure 5.6 shows the RX demodulator I/Q gain imbalance auto-calibration measurement result. On time slot 1, the circuits perform VGA gain lock, a large enough input voltage swing from the detector is needed during the gain imbalance calibration.

On the beginning of time slot 2, the I and Q path gains without auto-calibration were shown. We can see that the I signal output swing is larger than Q. During the calibration, the Q output swing become larger and larger, finally the calibration stops when Q output swing is larger than I. After the gain imbalance calibration, the gain mismatch is less than 0.1dB

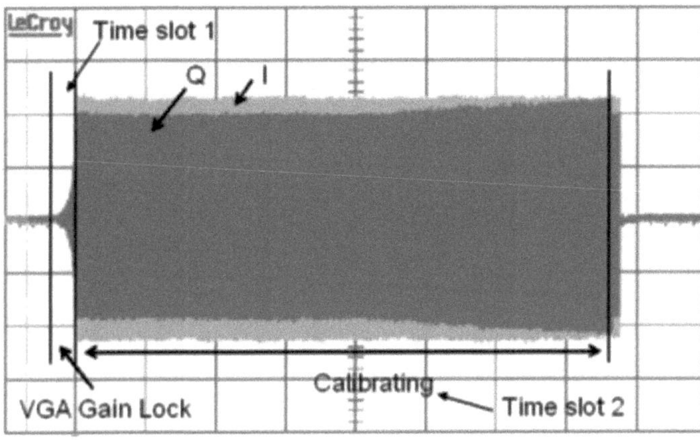

Figure 5-6. Measurement Result of RX I/Q Gain Imbalance Auto-Calibration.

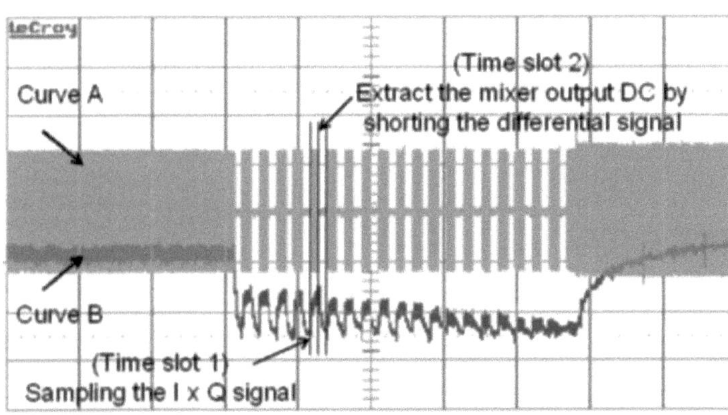

Figure 5-7. Measurement Result of Quadrature Phase Mismatch Auto-Calibration.

Figure 5.7 shows the RX demodulator quadrature phase mismatch auto-calibration measurement result. In the figure, curve A is the I path signal output, curve B is the RC low-pass filter output. On time slot 1, we sample the I mixing Q signal, on time slot 2, we extract the output DC voltage of the mixer by shorting the differential input signal. The difference between these two DC voltages is equal to Eq. (5.8). The Control Logic circuits are comparing these two DC voltages to determine appreciated control codes to the delay cell. From Figure 5.5, the difference between these two DC voltages becomes smaller and smaller, finally the S/C comparator output changes the polarity and calibration stops. Less than 1° quadrature phase error can be achieved after the calibration.

Chapter 6

SYSTEM MEASUREMENT RESULT

A fully integrated 802.11g transceiver IC is fabricated in a 0.25µm 1P5M CMOS technology. Its die microphotograph is shown in Figure 6.1 and typical measurement results of the transmitter are summarized in Section 6.1 and the results of the receiver are summarized in Section 6.2. The total die area is 10.2mm^2 (3mm x 3.4mm) and is housed in a 48-pin QFN package. A Printed Circuit Board (PCB) test board is shown in Figure 6.2; all the characteristic results are measured from this PCB.

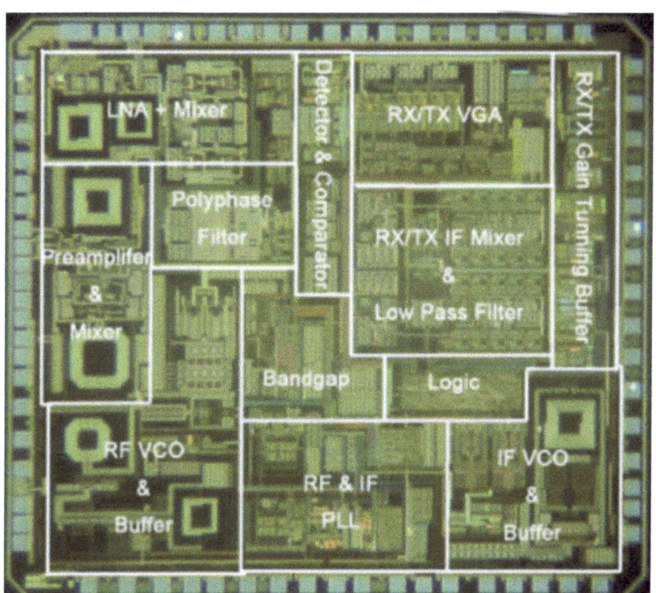

Figure 6-1. Die Micrograph (Data Provided by Muchip).

Figure 6-2. AC Characteristic Test Board (Data Provided by Muchip).

1. TRANSMITTER MEASUREMENT RESULT

Figure 6.3 shows the IF output channel power vs VGA Control voltage, where the CCK QPSK modulation signal is used in the test setup. Without degrading the signal EVM, our chip can deliver -7dBm (about -12.5dBm OFDM signal) modulation signal at the IF output. It can meet the system specification (to deliver -16dBm OFDM signal on the IF output) as discussed in Section 2.6.1. The gain range is about 20dB when control voltage varies from 1.4V to 2V.

Figure 6.4 shows the phase noise plots, the integration phase noise of the local signal is about 1.7°. Table 6.1 summaries the transmitter performance. Figure 6.5 shows the RF output spectrum and Figure 6.6 shows the RF output signal constellation under the 54Mbps data transmission mode. All the measurement results can meet the design goals.

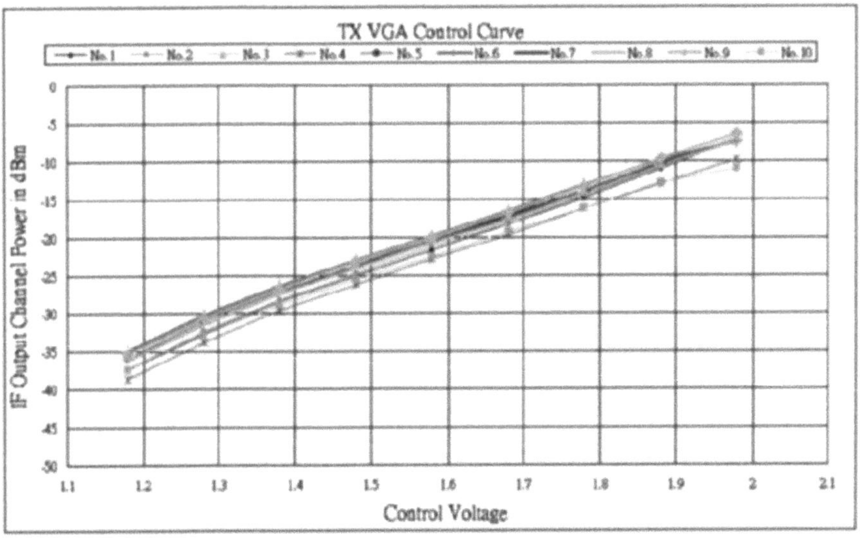

Figure 6-3. IF Output Power vs VGA Control Voltage.

Table 6-1. Transmitter Performance Summary

Items	Specification	Measurement
Supply Voltage	2.7 ~ 3.3	2.7 ~ 3.3
Supply Current	120mA	120mA
Baseband Input Voltage Gain Compression V_{1dB}	1V	1V
Max IF output power (OFDM signal)	-16dBm	-11dBm (Figure 6.3)
Single-Side Band Rejection	-40dBc	-44dBc (Figure 4.15)
Carrier Leakage	-25dBc	> 40dBc (Figure 4.15)
RF Output Gain Compression OP_{1dB}	6dBm	7dBm
IF/RF Section Gain	21dB	23dB
I/Q Gain Imbalance	0.1dB	0.1dB
I/Q Quadrature Phase Error	1°	1°

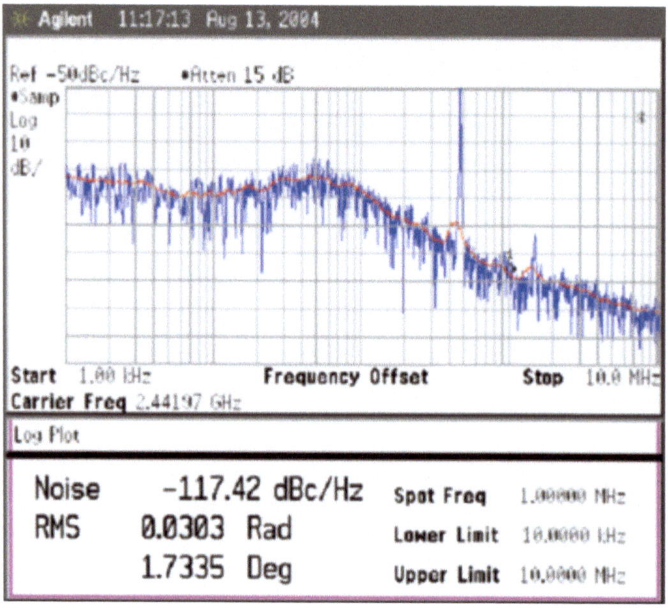

Figure 6-4. Phase Noise Plot (Data Provided by Muchip).

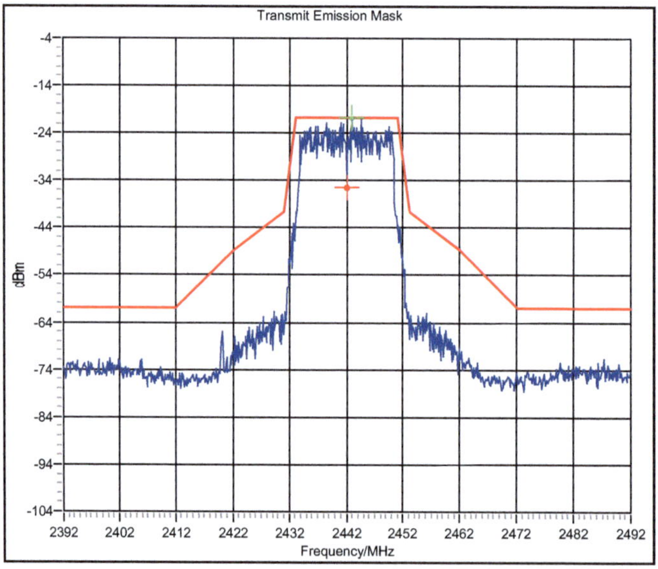

Figure 6-5. RF Output Spectrum Mask (Data Provided by Muchip).

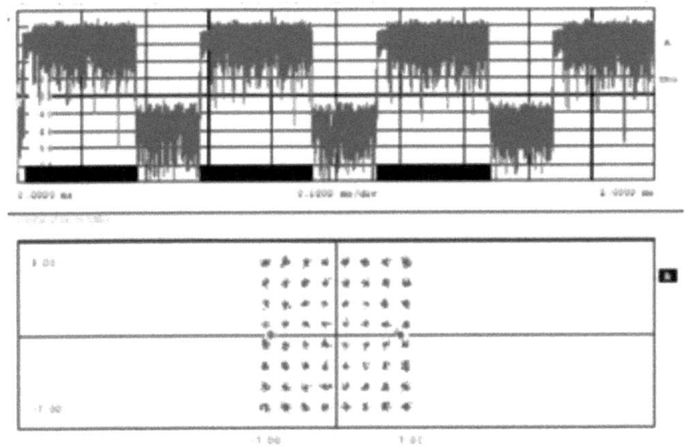

Figure 6-6. Constellation in a 802.11g 54Mbps Data Rate Mode (Data Provided by Muchip).

2. RECEIVER MEASUREMENT RESULT

Figure 6.7 shows the RF/IF section cascade noise figure and gain performance in high gain mode. The cascade IF and demodulator gain under control voltage varying from 1.1V to 2V is shown in Figure 6.8. Figure 6.9 shows the RX I and Q output swings. The measurement result of signal to noise ratio (SNR) vs input power is plotted in Figure 6.10, the picture overlaps the predicted results from Figure 2.14 for comparison. The maximum SNR in the measurement results is about 27.5dB; it is lower than our expectation. The limitation on maximum SNR is the poor IF/BB section cascade noise figure, but the performance still can meet the system performance requirement. The integration phase noise is shown in Figure 6.4 and the performance closes to our design goal. Table 6.2 gives a brief summary of the overall receiver performance. From the performance summary, the measurement results are almost the same as the design specification.

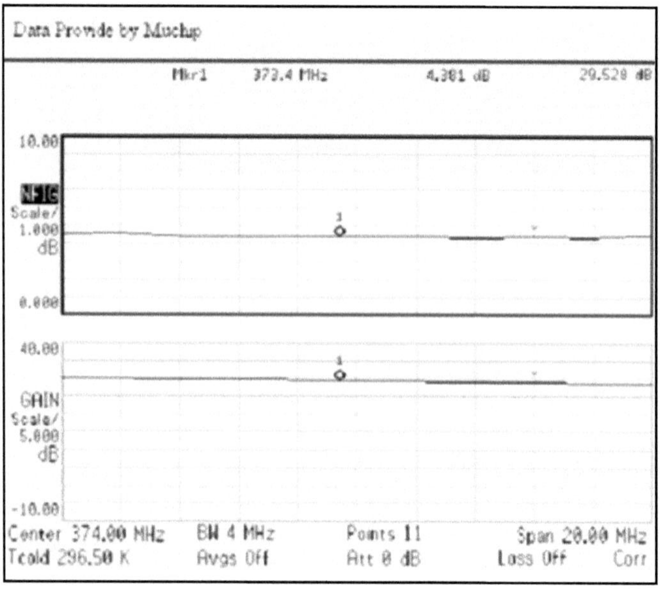

Figure 6-7. RF/IF Section Cascade Noise Figure and Gain.

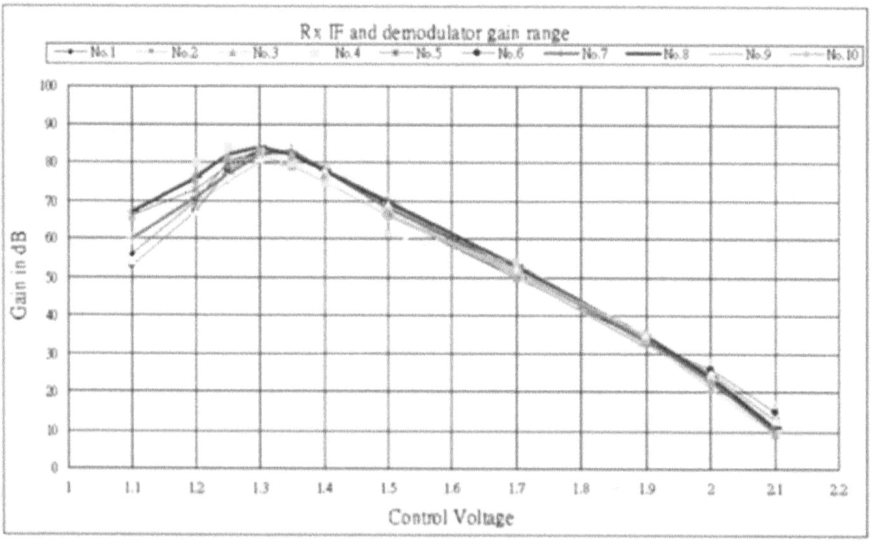

Figure 6-8. Cascade Gain of IF VGA and Demodulator under Different VGA Control Voltages.

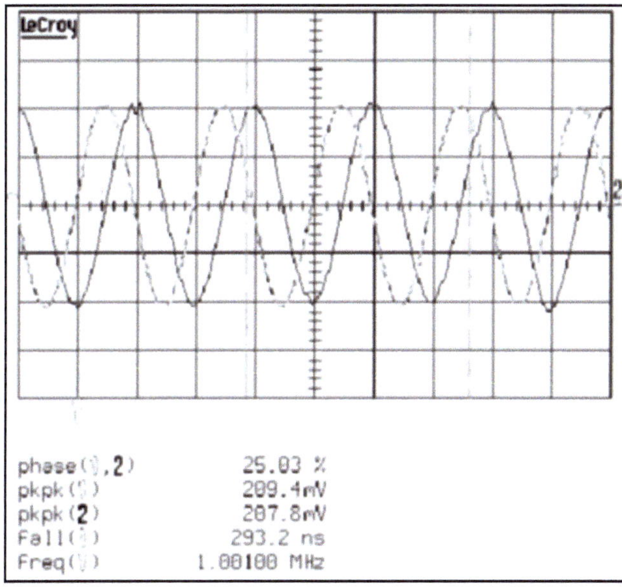

Figure 6-9. RX I and Q Output Voltage Swings (Less than 1° Quadrature Phase Error and 0.1dB Gain Imbalance).

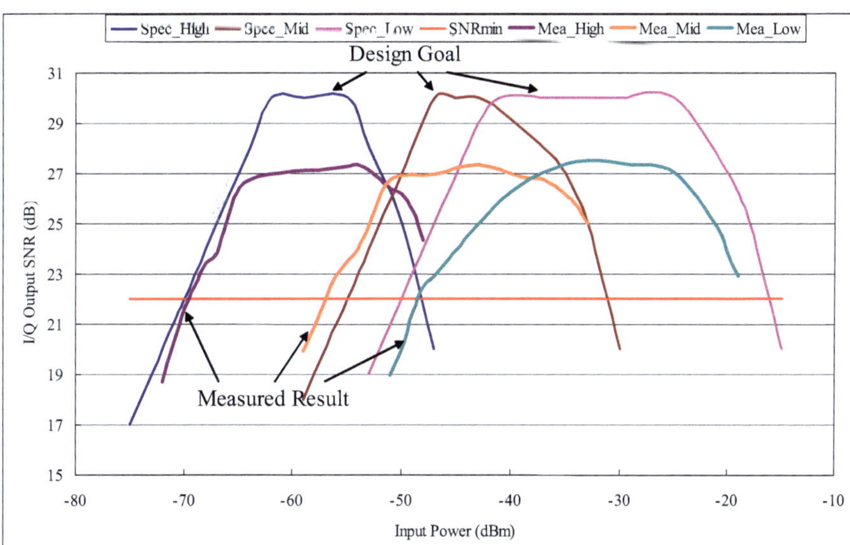

Figure 6-10. RX Output SNR vs RF Input Power.

Table 6-2. Receiver Performance Summary

Items		Specification	Measurement
Supply Voltage		2.7 ~ 3.3	2.7 ~ 3.3
Supply Current		150mA	150mA
RX FE Gain	High Gain Mode	31dB	29.5dB
	Middle Gain Mode	16dB	18dB
	Low Gain Mode	5dB	4.7dB
RX FE NF	High Gain Mode	4dB	4.4dB
	Middle Gain Mode	8dB	7.3dB
	Low Gain Mode	23.4dB	19dB
RX FE IP_{1dB}	High Gain Mode	-36	-35
	Middle Gain Mode	-25	-24.5
	Low Gain Mode	-12	-14
IF Gain Range (VGA Control from 1.1V to 2V)		16 ~ 80dB	15 ~ 80dB
Baseband Output Voltage Gain Compression V_{1dB}		1V	1V
I/Q differential output DC offset		15mV	15mV
I/Q Gain Imbalance		0.1dB	0.1dB
I/Q Quadrature Phase Error		1°	1°

Chapter 7

CONCLUSION

In the market of wireless local area network (WLAN), high data-rate transmission has been the trend. IEEE 802.11g standard works at 2.4GHz ISM band and supports data rates up to 54Mbits/s using OFDM modulation. To achieve the desired performance upper bound, it is crucial to deliver a close match of I and Q signals in both the modulator and the demodulator.

A CMOS IEEE 802.11g transceiver with an auto-I/Q calibration in TX modulator and RX demodulator, including DC cancellation, gain imbalance and quadrature phase error calibration are presented in this book. This is the first innovative transceiver architecture with a full I/Q calibration function. With the auto-I/Q calibration, the transceiver achieves a 0.1dB gain imbalance and a 1° quadrature phase error in both RX and TX. With the TX auto dc offset cancellation, the local leakage is less than 40dBc. The ideal of this architecture was published in International Solid-State Circuit Conference (ISSCC) 2005 [9]. The transceiver IC is fabricated in a 0.25μm 1P5M CMOS technology. The total die area is $10.2mm^2$ (3mm × 3.4mm) and is housed in a 48-pin QFN package.

This work also presents the IF/BB section circuits reuse in RX and TX. In Chapter 3, we show that after the well design of an IF VGA and a baseband low pass filter, these two circuits can be re-used in signal transmitting and receiving. Significant die area can be saved from this kind of circuits reuse.

Chapter 6 shows the performance summary of the transceiver; all the measurement results can meet the original transceiver system design goal. The performance of this chip has also been proved by the system link with Baseband processor, the data throughput of the whole system in higher than 22Mbps.

Future work may extend the proposed auto-I/Q calibration function to zero-IF and low-IF applications. Those architectures have the advantage on integration, but more tight performance requirements on calibration circuits have to be achieved.

References

[1] Wireless (Wi-Fi) Networking Equipment for Enterprises and Service Providers, Available: http://www.proxim.com/

[2] IEEE Wireless LAN Standards, Available: http://standard-s.ieee.org/getieee802/802.11.html

[3] C. Muschallik, "Influence of RF Oscillators on an OFDM Signal," *IEEE Transactions on Consumer Electronics,* Vol. 41, no. 3, pp. 592 – 603, Aug. 1995.

[4] D. Petrovic, W. Rave, and G. Fettweis, "Performance Degradation of Coded-OFDM due to Phase Noise," *in Proc. IEEE Vehicular Technology Conference*, Apr. 2003, Vol. 2, pp. 1168 – 1172.

[5] W. Eberle, J. Tubbax, B. Come, S. Donnay, H. De Man, and G. Gielen, "OFDM-WLAN Receiver Performance Improvement using Digital Compensation Techniques," *in Proc. IEEE Radio and Wireless Conference, RAWCON*, Aug. 2002, pp. 111-114.

[6] K. Vavelidis, I. Vassiliou, T. Georgantas, A. Yamanaka, S. Kavadias, G. Kamoulakos, C. Kapnistis, Y. Kokolakis, A. Kyranas, P. Merakos, I. Bouras, S. Bouras, S. Plevridis, and N. Haralabidis, "A Dual-Band 5.15-5.35 GHz, 2.4-2.5 GHz 0.18um CMOS Transceiver for 802.11 a/b/g Wireless LAN," *IEEE J. Solid-State Circuits,* vol. 39, no. 7, pp. 1180-1184, July 2004.

[7] M. Zannoth, T. Ruhlicke, and B.-U. Klepser, "A Highly Integrated Dual-Band Multimode Wireless LAN Transceiver," *IEEE J. Solid-State Circuits,* vol. 39, no. 7, pp. 1191-1195, July 2004.

[8] R. Ahola, A. Aktas, J. Wilson, K.R. Rao, F. Jonsson, I. Hyyrylainen, A. Brolin, T. Hakala, A. Friman, T. Makiniemi, J. Hanze, M. Sanden, D. Wallner, Y. Guo, T. Lagerstam, L. Noguer, T. Knuuttila, P. Olofsson, and M. Ismail, "A Single-Chip CMOS Transceiver for 802.11a/b/g Wireless LANs," *IEEE J. Solid-State Circuits,* vol. 39, no. 12, pp. 2250-2258, Dec 2004.

[9] Y.H. Hsieh, W.Y. Hu, S.M. Lin, C.L. Chen, W.K. Li, S.J. Chen, and D.J. Chen, "An Auto I/Q-Calibrated CMOS Transceiver for 802.11g," *in Proc. ISSCC,* Feb, 2005, pp. 92-93.

[10] A. Georgiadis, "Gain, Phase Imbalance, and Phase Noise Effects on Error Vector Magnitude," *IEEE Transactions on Vehicular Technology*, vol. 53, no. 2, pp. 443-449, March 2004.

[11] B. Razavi, *RF Microelectronics*. Englewood Cliffs, NJ: Prentice Hall, 1998.

[12] A. Zolfaghari, A. Chan, and B. Razavi, "Stacked Inductors and Transformers in CMOS Technology," *IEEE J. Solid-State Circuits,* vol. 36, no. 4, pp. 620-628, April 2001.

[13] H. Samavati, H.R. Rategh, and T.H. Lee, "A 5-GHz CMOS Wireless LAN receiver Front End," *IEEE J. Solid-State Circuits,* vol. 35, no. 5, pp. 765-772, May 2000.

[14] J.W.M. Rogers and C. Plett, "A 5-GHz Radio Front-End With Automatically Q-Tuned Notch Filter and VCO," *IEEE J. Solid-State Circuits,* vol. 38, no. 9, pp. 1547-1554, September 2003.

[15] SAWTEK 855898 Data Sheet, Available: http://www.sawtek.com/

[16] M. Zargari, D.K. Su, C.P. Yue, S. Rabii, D. Weber, B.J. Kaczynski, S.S. Mehta, K. Singh, S. Mendis, and B.A. Wooley, "A 5-GHz CMOS Transceiver for IEEE 802.11a Wireless LAN Systems," *IEEE J. Solid-State Circuits,* vol. 37, no. 12, pp.1688-1694, Dec 2002.

[17] P.C. Huang, L.Y. Chiou, and C.K. Wang, "A 3.3V CMOS Wideband Exponential Control Variable Gain Amplifier," *in Proc. IEEE Int. Symposium on Circuit and Systems*, Montery, Jun. 1998, vol. 1, pp. 285-288.

[18] A.M. Ismail and A.M. Soliman, "Novel CMOS Wide-Linear-Range Transconductance Amplifier," *IEEE Transactions on Circuit and System-I: Fundamental Theory and Application,* vol. 47, no. 8, pp. 1248-1253, August 2000.

[19] B. Razavi, "A 5.2-GHz CMOS Receiver with 62-dB Image Rejection," *IEEE J. Solid-State Circuits,* vol. 36, no. 5, pp. 810-815, May 2001.

[20] B. Razavi, *CMOS Analog Circuit Design*, Englewood Cliffs, NJ: Prentice Hall, 1998.

[21] M. Fualkner, T. Mattsson, and W. Yates, "Automatic Adjustment of Quadrature Modulators," *Electronics Letters,* vol. 27, no. 3, pp. 214-216, Jan 1991.

[22] L. Der and B. Razavi, "A 2-GHz CMOS Image Reject Receiver with LMS Calibration," *IEEE J. Solid-State Circuits,* vol. 38, no. 2, pp. 167-175, Feb 2003.

[23] S.P. Wu and Y. Bar-Ness, "OFDM Systems in the Presence of Phase Noise: Consequences and Solutions," *IEEE Transactions on Communications,* vol. 52, no. 11, pp. 1988-1996, Nov 2004.

[24] IEEE 802.11g White Paper, Available: http://www.broadcom.com/collateral/wp/802. 11g- WP104-R.pdf

MIX
Papier aus verantwortungsvollen Quellen
Paper from responsible sources
FSC® C105338

If you have any concerns about our products,
you can contact us on
ProductSafety@springernature.com

In case Publisher is established outside the EU,
the EU authorized representative is:
Springer Nature Customer Service Center GmbH
Europaplatz 3, 69115 Heidelberg, Germany

Printed by Libri Plureos GmbH
in Hamburg, Germany